PERGAMON INTERNATIONAL LIBR
of Science, Technology, Engineering and Social Studies

The 1000-volume original paperback library in aid of education,
industrial training and the enjoyment of leisure

Publisher: Robert Maxwell, M.C.

The Historical Supernovae

THE PERGAMON TEXTBOOK
INSPECTION COPY SERVICE

An inspection copy of any book published in the Pergamon International Library
will gladly be sent to academic staff without obligation for their consideration for
course adoption or recommendation. Copies may be retained for a period of 60 days
from receipt and returned if not suitable. When a particular title is adopted or
recommended for adoption for class use and the recommendation results in a sale
of 12 or more copies the inspection copy may be retained with our compliments.
The Publishers will be pleased to receive suggestions for revised editions and new
titles to be published in this important international Library.

Other Titles of Interest:

BEER:	Vistas in Astronomy (review journal)
BOWLER:	Gravitation and Relativity
ELGAROY:	Solar Noise Storms
GLASBY:	The Nebular Variables
HEY:	The Radio Universe , 2nd edition
HILLAS:	Cosmic Rays
MEADOWS:	Early Solar Physics
PACHOLCZYK:	Radio Galaxies
REDDISH:	Stellar Formation
ROWAN-ROBINSON:	Far Infrared Astronomy

The Historical Supernovae

BY

DAVID H. CLARK
Mullard Space Science Laboratory
University College London

AND

F. RICHARD STEPHENSON
Institute of Lunar and Planetary Sciences
University of Newcastle upon Tyne

PERGAMON PRESS

OXFORD · NEW YORK · TORONTO · SYDNEY · PARIS · FRANKFURT

U.K.	Pergamon Press Ltd., Headington Hill Hall, Oxford OX3 0BW, England
U.S.A.	Pergamon Press Inc., Maxwell House, Fairview Park, Elmsford, New York 10523, U.S.A.
CANADA	Pergamon of Canada Ltd., 75 The East Mall, Toronto, Ontario, Canada
AUSTRALIA	Pergamon Press (Aust.) Pty. Ltd., 19a Boundary Street, Rushcutters Bay, N.S.W. 2011, Australia
FRANCE	Pergamon Press SARL, 24 rue des Ecoles, 75240 Paris, Cedex 05, France
WEST GERMANY	Pergamon Press GmbH, 6242 Kronberg-Taunus, Pferdstrasse 1, Frankfurt-am-Main, West Germany

First edition 1977

Library of Congress Cataloging in Publication Data

Clark, David H
The historical supernovae.

1. Supernovae. 2. Stars--Observations--History.
I. Stephenson, Francis Richard, Joint author. II. Title.
QB841.C58 1977   532.8'446   76-44364
ISBN 0-08-020914-9 (Hardcover)
    0-08-021639-0 (Flexicover)

In order to make this volume available as economically and rapidly as possible the author's typescript has been reproduced in its original form. This method unfortunately has its typographical limitations but it is hoped that they in no way distract the reader.

Printed in Great Britain by A. Wheaton & Co., Exeter

To Suzanne and Ellen

"The investigation of the remnants of supernovae
and their relation to historical records, both
written and unwritten, will be one of the most
fascinating tasks awaiting the next generation
of astronomers ......"

Fritz Zwicky - joint instigator of the term
supernova, and a pioneer of modern supernova
surveys;
(in "Stellar Structure", Volume VIII of "Stars
and Stellar Systems", Ed. Aller and McLauglin.
University of Chicago Press. 1965)

"The extent to which the Chinese records of guest
stars remain of living interest to current
astronomical research may be seen in the field
of radio-astronomy, where during the past few
years great additions to knowledge have been
made......The rapid upsurge of this new and
powerful method of study of the birth and death
of stars....makes urgently necessary the reduction of the information contained in the ancient
and medieval Chinese texts to a form utilisable
by modern astronomers in all lands. For this
purpose, however, collaboration between competent
sinologists and practical astronomers and radio
astronomers is indispensable."

Joseph Needham, F.R.S. - distinguished historian
of Chinese Science;
(in "Mathematics and the Sciences of the Heavens
and the Earth", Volume 3 of "Science and Civilisation in China". Cambridge University Press.
1959)

CONTENTS

Preface ix

Chapter 1. New Stars – Novae and Supernovae 1

Chapter 2. The Search for the Historical Records of Supernovae 14

Chapter 3. A Classification of the Far-Eastern New Stars 40

Chapter 4. The Search for the Remnants of Supernovae 57

Chapter 5. The Guest Star within the Southern Gate 83

Chapter 6. The Chin Dynasty Guest Stars 102

Chapter 7. An Extensively observed New Star of Extreme Brightness 114

Chapter 8. The Birth of the Crab Nebula 140

Chapter 9. The Invader of the Guest Houses 161

Chapter 10. The New Star of Tycho Brahe 172

Chapter 11. The New Star of Johannes Kepler 191

Chapter 12. Some thoughts on the Evolution of Supernova Remnants 207

References 223

Index 229

PREFACE

In writing this book it has been our intention to approach the subject of supernovae by way of the historical records, with particular reference to the history of the Far East. The book therefore is an interdisciplinary study suitable for students of Astronomy, History of Science or Sinology, as well as the non-specialist.

When directing a work to such a wide audience, clearly allowance must be made for the varying backgrounds and interests of readers. Consequently Chapters 1 to 4 contain much introductory material to assist in the understanding of the remaining eight chapters.

The impact of the study of historical supernovae on modern astrophysics has been very significant. Readers need only be reminded that many of the most dramatic discoveries in Astronomy during the past few years, such as pulsars, black holes, X-ray binaries, etc., are directly linked with supernova explosions. We have not hesitated to describe the historical records of supernovae as "among the most valuable legacies which the ancient world has bequeathed to modern science".

During the preparation of this book we have benefited much from collaboration with Mr. A. C. Barnes, School of Oriental Studies, University of Durham, and Dr. J. L. Caswell, Radiophysics Division, CSIRO, Australia. It is a pleasure to acknowledge our gratitude to both colleagues. We are indebted to Professor A. J. Meadows, Department of Astronomy and History of Science, University of Leicester, for reading the manuscript and offering several valuable criticisms and suggestions, and also to Professor W. H. McCrea, Astronomy Centre, University of Sussex, for collaboration on some of the material in Chapter 12.

Our special thanks are due to Miss Jane Salton for typing the manuscript. The help of Dr. T. J. Saunders and Dr. A. J. Woodman, both of the Department of Classics, University of Newcastle upon Tyne, with difficult Latin texts is also gratefully acknowledged.

Finally we wish to thank our families for their continued encouragement during the preparation of this book.

David H. Clark

F. Richard Stephenson

November, 1976.

ACKNOWLEDGEMENTS

We are grateful to the following for permission to publish copyright material -

The British Museum, London - Plate 1.
The Royal Scottish Museum, Edinburgh - Plate 5.
University of Chicago Press - Plates 6, 7 and 9 are from "An Optical Atlas of Galactic Supernova Remnants" by van den Bergh, Marscher, and Terzian, The Astrophysical Journal Supplement No. 227. © 1973 by the American Astronomical Society. All rights reserved.

Chapter 1

NEW STARS — NOVAE AND SUPERNOVAE

The vast majority of stars in a typical galaxy such as our own
(the Milky Way) are extremely stable, and produce a remarkably
steady output of radiation over millions of years. In marked
contrast are the Novae and Supernovae - stars which spontan-
eously explode with a spectacular and rapid increase in brightness, so that at maximum they rival the brightest stars in the
heavens, before gradually fading into insignificance. At its
height a nova may be intrinsically a million times as bright as
the Sun, with a supernova at its height being several thousand
times as bright again and emitting as much energy as all the
other stars in the galaxy combined. The term "nova" means literally a new star, and was introduced for such objects before
the realisation that a star actually existed at the site of the
nova or supernova prior to the outburst.

A supernova is now recognised as the violent end to the evolution of a certain type of star. Although the exact reason why
some stars undergo such catastrophic self-destruction is not
fully understood, the evolutionary path leading up to at least
one type of supernova event may be described quite simply. As
a nascent star forms from a cloud of gas and dust collapsing
under the action of gravitational forces, the temperature in the
interior of the cloud increases to the point ($\simeq 10^7$ $^{\circ}$K) where
certain nuclear reactions can take place. At first the dominant
nuclear reaction is the so-called proton-proton chain reaction,
the net effect of which is the fusion of hydrogen to form helium
with the release of energy. The liberation of this thermonuclear energy halts the gravitational contraction, and the star
settles down to the relatively stable state in which it spends
most of its active life - "burning" hydrogen in the central core
to balance the energy radiated from the surface.

Although the nuclear fuel reserves of a star are enormous, they
are not unlimited. When the core hydrogen is expended, a
renewed contraction of the core raises the temperature to the
level ($\simeq 2 \times 10^8$ $^{\circ}$K) required to initiate the burning of helium
to form carbon and oxygen, accompanied by hydrogen burning in
the surrounding shell. Later stages follow involving the fusion
of successively heavier elements all the way to iron, when no
further reactions can extract energy. Since the energy yield
per fusion and the abundance of the reacting elements both
decrease at each successive stage of nuclear burning, the star
squanders its remaining fuel reserves at an ever increasing rate
as it approaches its final destruction. Several stages of nuclear burning occur simultaneously in the same star, with the
core being most advanced and reaching the iron end-point while

the outer portion of the stellar envelope is still burning
hydrogen, with intervening zones at intermediate stages.

What follows when no further nuclear energy is available in the
stellar core depends on the mass of the star. Those stars with
core masses less than about 1.4 $M_\odot$ ($M_\odot \equiv$ the mass of our Sun)
undergo a relatively controlled collapse to a state of very high
density ($\simeq 10^6$ grams per cm^3), forming a so-called white dwarf.
By contrast, stars with cores heavier than 1.4 $M_\odot$ experience
rapid collapse accompanied by the release of gravitational
energy. For particularly massive stars the core mass and its
kinetic energy of implosion may drive the system beyond nuclear
densities to complete gravitational collapse to a black hole.
However the collapse of the cores of less massive imploding
stars (those with masses in the range 1.4 to. 3.5 $M_\odot$) will be
halted at nuclear density to give a rapidly rotating neutron
star (density $\simeq 10^{14}$ grams per cm^3), which may be observable as
a pulsar emitting pulses of radiation over a wide range of
frequencies at intervals of a fraction of a second. The sudden
stopping of the collapsing core at nuclear density yields an
enormous temperature ($\simeq 10^{12}$ $^\circ$K) and high pressure, producing
an outward moving shock wave. The stellar envelope, still
containing elements lighter than iron from intermediate stages
of nuclear burning, and collapsing behind the core, will suddenly experience this shock with resulting rapid compression and
sharp increase in temperature. The nuclear reaction involving
the fusion of oxygen-16 nuclei to form silicon (at a temperature
of about 10^9 $^\circ$K) is potentially explosive, and the initiation
of this reaction in the collapsing envelope is believed to
contribute to the explosive ejection of a significant fraction
of the stellar mass. These final stages of the evolution of a
massive star are depicted schematically in Fig. 1.1.

A supernova event thus results in the sudden implosion of the
core (on a time-scale of seconds) to leave a rapidly rotating
collapsed stellar remnant, the explosive ejection of the stellar
envelope broken up into filamentary structure flying off with
velocities of thousands of kilometres per second, and the
release of vast quantities of energy of the order of 10^{51} ergs
(equivalent to about 10^{28} 1-Megaton hydrogen bomb explosions).

Supernovae are of tremendous interest to astrophysicists, not
only because they represent the most spectacular of stellar
events, but because the remnants and ejecta of such explosions
are amongst the most unusual and exciting of astrophysical
objects and phenomena. In addition to pulsars and black holes,
supernova explosions are believed to be responsible for the
production of high-velocity "runaway" stars hurtling through
the galaxy at speeds approaching a million miles per hour, the
high-energy cosmic rays continually bombarding our planet, the
heavy elements, spectacular expanding nebulosities which are
amongst the most beautiful objects in the heavens, extended
sources of radio emission, possibly gravitational radiation,
and probably most of the galactic X-ray sources.

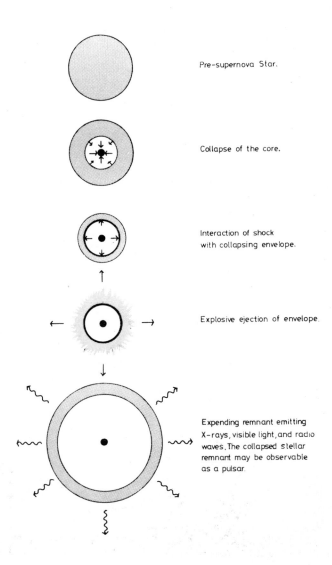

Pre-supernova Star.

Collapse of the core.

Interaction of shock
with collapsing envelope.

Explosive ejection of envelope.

Expanding remnant emitting
X-rays, visible light, and radio
waves. The collapsed stellar
remnant may be observable
as a pulsar.

Fig. 1.1.   A schematic representation of the self-destruction
of a star in a supernova event.

By contrast with a supernova, which is a "one-off" event herald-
ing the final destruction of a star, a nova is believed to be
merely a temporary departure of certain stars from the well-
established stellar evolutionary path, and may in fact be a rep-
etitive phenomenon with the interval between outbursts being up
to several thousand years. In each nova explosion only a very
small fraction of the star's mass (possibly no more than one
hundredth of one percent) is believed to be ejected, and some of
this may eventually be drawn back onto the star. Although the
exact reason why certain stars experience nova explosions remains
open to debate, the most widely held theory is based on the now
convincing observational evidence that a prenova is a <u>close bin-
ary</u> system. A close binary system is a pair of stars which are
close enough for mass transfer between them to occur at certain
times during the normal course of their evolution. Any such
system containing a white dwarf and a normal companion star is a
potential nova. During an expansion phase of the companion star,
material (mainly hydrogen) is transferred to the atmosphere of
the white dwarf, where under the action of the strong gravitational field it is compressed and heated to the ignition temperature for hydrogen burning reactions. Thereafter thermonuclear
runaway could produce the explosive ejection of matter and the
radiation of energy seen as the nova event. The interchange of
material within a close binary system preceding a nova outburst
is depicted in Fig. 1.2.

A nova explosion resembles a supernova in many respects, but is
on a much smaller scale. A typical nova lasts days or at most
weeks rather than months, and the energy released is less than
one ten-thousandth of that in a supernova explosion. The expanding shell of gas from the explosion dissipates in a few decades,
leaving the binary system showing little change from its original

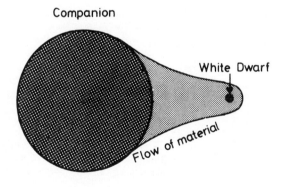

Fig. 1.2. The interchange of material within a
close binary system preceding a nova outburst.

condition. Novae are much more frequent than supernovae, and for a short time a nearby nova can shine among the brightest stars in the sky. There have been six bright novae in the present century, appearing in 1901, 1918, 1925, 1934, 1942 and 1975. All temporarily rivalled first magnitude stars, and for a few days the 1918 nova was almost as bright as Sirius (the brightest permanent star in the sky).

Although novae and supernovae differ in their origin, in the scale of outburst, and indeed in their frequency of occurrence, the existence of supernovae as events distinct from common novae has been recognised only in the last forty years.

There have been many outbursts of an apparently stellar nature recorded throughout history. Most early observations of "new stars" were made in the Far East - China, Japan and Korea. In ancient and medieval Europe and the Arab Lands there seems to have been little interest in such phenomena, partly due to the widespread influence of the Aristotelian doctrine of a perfect changeless celestial sphere and partly the result of sheer inability to recognise a new star-like object (see Chapter 2). This situation was not to change until the 16th century.

In the Far East, professional astronomers/astrologers were employed by the ruler to maintain a constant watch of the sky, and to report and interpret any unusual events which might happen. Many such events are reported in the astronomical treatises of the various dynastic histories. We have remarkably detailed records from China going back to about 200 BC, but unfortunately, probably as the result of the famous "Burning of the Books" at the direction of Ch'in Shih-huang, the first true emperor of China, in 213 BC, there is very little before that date. In Japan and Korea, regular astronomical observations began around AD 1000, so that after this date we find frequent duplication. The historical background of the Far Eastern observations will be further discussed in Chapter 2.

Ideally, what we would like to find among the historical records are accounts of new stars describing in detail the position, changes in brightness, colour and duration of visibility. Such completeness is seldom realised in practice, but it is important to take full advantage of whatever information we can obtain from the early observations. Several of the historical records of new stars make reference to extreme brightness. However it is not possible to make a distinction between novae and supernovae in our own galaxy on this basis alone, since the apparent luminosity of novae occurring close to the Earth would not differ from that of the intrinsically very much brighter supernovae at greater distance.

There were two principal reasons for originally suggesting that there existed stellar explosions thousands of times as bright as common novae. The first of these was based on the observations made of new stars in extragalactic stellar systems. In 1885 a temporary star was observed near the nucleus of Messier 31 which nearly equalled the apparent brightness of that galaxy, and since

that time several hundred new stars have been detected in external galaxies. If the number of such events in different apparent
brightness ranges is plotted for a few nearby galaxies, it
immediately becomes apparent that there is a small, but distinct,
group of extreme brightness (see Fig. 1.3). (The scale of
stellar magnitudes is a logarithmic one in which negative numbers
indicate a greater brightness than positive numbers, and in which
each magnitude is some 2.5 times brighter than the one below it.
The observed magnitude of a star is called its <u>apparent</u> magnitude.
For example, zero apparent magnitude is the brightness of Vega,
and -4 apparent magnitude is the brightness of Venus. To compare
one star with another we need to know their intrinsic magnitudes,
such as would be obtained if the stars were placed at a standard
distance. The distance chosen is 32.6 light years, = 10pc, and
the magnitude at this distance is termed the <u>absolute</u> magnitude.)

The second reason for suggesting the existence of supernovae
was related to the new star of AD 1572, described in some detail
and accurately positioned by the eminent Danish astronomer Tycho
Brahe. Ordinary novae show an increase from pre-nova to maximum
brightness typically of about 15 magnitudes. The new star of
1572 was for a few days as bright as Venus, about apparent
magnitude -4. If an ordinary nova, it must have been within a
few tens of light years of the Earth, and its stellar remnant
should now not be fainter than +11. However no such object
could be found, and it was concluded that the new star of
AD 1572 must have been at large distance and with extreme intrinsic brightness at maximum.

Thus, in 1937, the term <u>supernova</u> was proposed for the distinct
group of new stars of extreme intrinsic brightness by Zwicky and
Baade, two of the pioneers of supernova research.

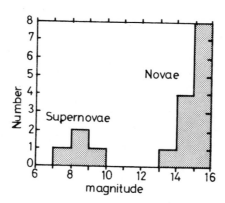

Fig. 1.3. A distribution of bright novae and
supernovae in nearby galaxies (from Zwicky 1965).

Three other galactic new stars recorded historically have for
some time been recognised to be supernovae. A new star which
appeared in the year AD 1604 in the constellation of Ophiuchus
was carefully observed for about 12 months by many of the leading astronomers of Europe and independently by the court astronomers of Korea and China. The celestial location of the star
was so accurately measured by Kepler and Fabricius (to better
than a minute of arc) that Baade (1943) had no difficulty in
locating a small patch of nebulosity which clearly represented
its optical remnant.

The new star of AD 1054 was recorded only by the Chinese and
Japanese. Combining the various reports, we learn that the star
was visible in daylight for a total of 23 days, with a total
period of visibility exceeding 22 months. That this event was
associated with the Crab Nebula was possibly first suggested by
Lundmark (1921) ; certainly the rate of expansion of the optical
filaments of the nebula gives a date of outburst in close
agreement with AD 1054, and there is every reason to believe that
the Crab Nebula is one of the few select supernova remnants of
precisely known age.

The new star which appeared in AD 1006 is the only such star
known to be recorded in European and Arabic history before the
Renaissance. It was also extensively observed by the Chinese
and Japanese astronomers. Many texts comment on the extreme
brilliance of the star. For example ; "It shone so brightly
that objects could be seen by its light" (China) ; "Its rays
on the Earth were like the rays of the Moon" (Iraq) ; "Its
light illuminated the horizon and ... its brightness was a little
more than a quarter of the brightness of the Moon" (Iraq). One
Chinese record states that the star was followed for several
years. Goldstein (1965) not unrealistically estimated the
apparent magnitude at maximum as between -8 and -10.

There can be little doubt that the long-duration new stars of
extreme brilliance sighted in AD 1006, AD 1054, AD 1572 and
AD 1604 were supernovae. (They will be discussed in detail in
later chapters.) At least four supernovae occurring on the near-
side of the galaxy during the past millennium would seem to
place a lower limit on the frequency of galactic supernovae of
about one per 120 years. However it must be stressed that the
short interval between some actual sightings of supernovae can
give a grossly misleading impression of the frequency of events
because of the extreme distances involved. For example, the
AD 1006 supernova occurred at a distance of about 3 thousand
light years from the Earth, while that of AD 1054 occurred at a
distance of over 6 thousand light years, so that although first
sighted on Earth within half a century of each other the outbursts were actually separated in time by over 3000 years.
Despite this cautionary note, there is growing evidence that in
galaxies such as our own supernovae occur, on average, about once
per century.

The fact that few supernovae have been recorded historically is

Fig. 1.4.   A schematic edge-on view of the Galaxy.

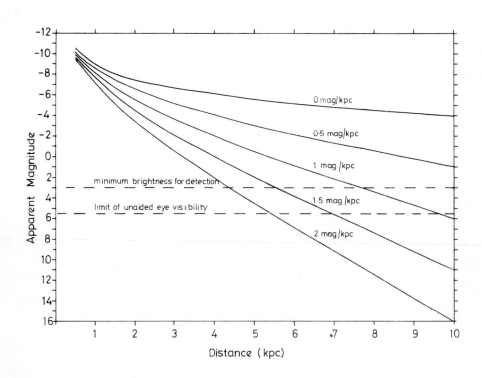

Fig. 1.5.   Curves of apparent (or observed) magnitude of a supernova (absolute magnitude -19), as a function of distance for selected values of interstellar absorption between 0 and 2 magnitudes per kpc.

partly a consequence of the high degree of optical obscuration
produced by dust clouds which permeate the whole galaxy and
blanket out many such events. A large portion of the Milky Way
lying towards the galactic centre (which is located in the
constellation of Sagittarius) is effectively screened from us
optically, The situation is not improved by the Sun's location
almost exactly on the galactic plane where the concentration of
dust is highest - see Fig. 1.4. (The distance scale in Fig. 1.4
is in kiloparsecs, kpc, where one parsec is the distance of a
star with an angle subtended by the radius of the Earth's orbit
of one second of arc. One parsec is equivalent to 3.263 light-
years.) In spite of its extreme intrinsic brilliance, a supernova in our galaxy is unlikely to be detected unless it lies
at a considerable distance to the north or south of the galactic
plane or else happens to occur in the Sun's neighbourhood. This
is highlighted in Fig. 1.5, which plots the diminishing apparent
magnitude for a supernova of maximum absolute magnitude -19,
with increasing distance from the Earth, and for selected values
of apparent interstellar absorption.

In contrast, external galaxies are visible to us from all directions, and because of their small angular size can be studied as
a unit. These are just two of the reasons why the extra-galact-
ic supernova searches at Mt. Palomar, Asiago (Italy), and elsewhere have been so successful. As many as three or even four
supernovae have been detected in each of a number of external
galaxies in the past few decades, whereas none has been seen in
our own galaxy since AD 1604, 5 years before Galileo Galilei
first applied the telescope to astronomical research.

The extragalactic surveys have provided important new insights
into the nature of supernovae. In particular, it has been
found that two main types of supernovae can be recognised on the
basis of absolute luminosities, light curves, and spectra.
The so-called Type I supernovae typically have an absolute
magnitude at maximum of -19, and after an initial drop of about
3 magnitudes in 20 to 30 days the light curve shows an approximately exponential decay (see Fig. 1.6.) Type II supernovae
are rather more individualistic, although a common feature of
their light curves is a "shoulder" after maximum, followed by a
rather rapid decline (see Fig. 1.7.) The composite supernova
light curves of Figs. 1.6 and 1.7 may be compared with those
for common novae shown in Fig. 1.8. Type I supernovae are
believed to be produced by stars of mass slightly greater than
one solar mass, while Type II supernovae are the progeny of very
much more massive stars and may involve the ejection of several
solar masses in the explosion and there are clear indications
that ejections at different velocities occur.

The maximum apparent magnitude for a supernova may be crudely
estimated from an historical record giving only the period of
visibility. As noted earlier, and depicted in Fig. 1.6, a
Type I supernova light-curve falls about 3 magnitudes in 20 to
30 days; the subsequent decline from maximum absolute magnitude
is approximately 4 magnitudes after 100 days and approximately

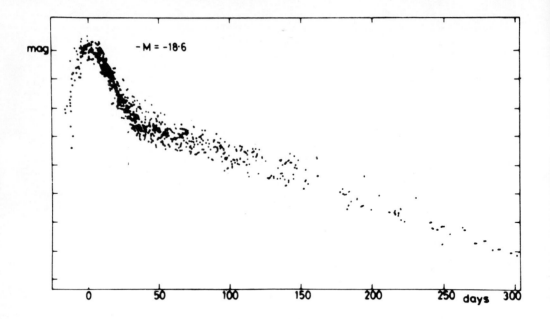

Fig. 1.6. Composite blue light curve obtained
by the fitting of the observations of 38 Type I
supernovae. One magnitude intervals are marked
on the ordinate. (From Barbon et al, 1974a).

7 magnitudes after 300 days, etc. The limit of naked-eye
visibility is about magnitude +5.5. Thus, for example, a recorded duration of visibility of 300 days would imply a maximum
absolute magnitude of -1.5 assuming typical Type I supernova
behaviour. The technique is of course highly subjective, and
cannot be used assuming Type II parameters because of the uncertain behaviour of Type II supernovae beyond about 100 days.
Nevertheless, we will have cause to apply the technique later
to give preliminary estimates for the brightness and thus the
distance of certain supernovae.

Lacking any telescopic observations of galactic supernovae,
it is not surprising that today astrophysicists are concentrat-
ing much effort on the study of supernova remnants (SNRs). A
supernova may produce a number of possible observable remnants ;
the main ones being a pulsar, an expanding optical nebulosity,
an extended radio source, and an extended X-ray source (see
Chapter 4). The shock wave expanding from the initial outburst
is usually found to delineate a region of radio emission, while
the heated interstellar material swept up by the shock wave may
act as an observable source of X-rays. Obscuration means that
only the nearby supernova remnants can be observed at optical and

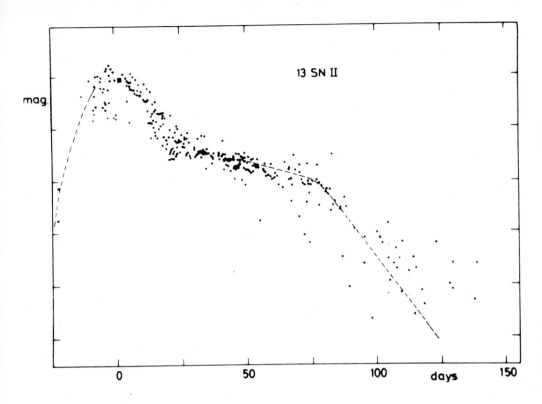

Fig. 1.7. Composite light curve obtained from
13 Type II supernovae. (From Barbon et al, 1974b).

X-ray wavelengths. However the radio remnants of supernovae
can be detected on the remote side of the galaxy.

Over 120 extended galactic radio sources have been identified as
SNRs. To date, no attempt to detect radio emission from extragalactic SNRs (other than 14 detected in the Magellanic Clouds)
have produced conclusive results. We have therefore the anomaly
of many extragalactic supernovae observed but no unambiguously
identified extragalactic SNRs, and at the same time no galactic
supernovae observed in modern times but many galactic SNRs.

Most of the catalogued galactic SNRs are probably so old that no
records of the supernova explosion which created them can be
expected to exist. For this reason, the few available historical observations of supernovae, although crude by present day
standards, are very important. If a particular object can be
identified with certainty as the remnant of a new star recorded
in a particular year, then the time that the remnant has been
developing is precisely known. Several such identifications

(a)

(b)

Fig. 1.8.   (a) The light curve of a 'fast' nova, showing the
typical rapid decrease in brightness after maximum - Nova Aquilae
(1918), the brightest discovered since the invention of the telescope.   (From Spencer Jones, 1961).   (b) The light curve of a
'slow' nova, showing the typical irregular variation of brightness - Nova Pictoris (1925).   (From Payne - Gaposchkin, 1957).

could then provide valuable observational evidence to test
current theories on the evolution of supernova remnants, and on
the nature of supernova explosions. Hence there can be little
doubt that the historical astronomical records must be regarded
as among the most valuable legacies which the ancient world has
bequeathed to modern science.

Chapter 2

THE SEARCH FOR THE HISTORICAL
RECORDS OF SUPERNOVAE

Usable historical records of new stars appear to be found in
only four principal sources : Medieval European monastic chronicles; Arabic chronicles, astrological works etc.; post
renaissance European scientific writings; and Far Eastern
histories and diaries.

Possible historical sources of new star observations which might
come readily to mind are the Greek and Roman Classics and the
Late Babylonian astronomical texts. However, these turn out to
be of little value. It is doubtful whether there are any useful
references to new stars in the Classics. According to Pliny
(Natural History, Book II) the appearance of a new star (nova
stella) inspired Hipparchus to compile the first known star
catalogue, giving co-ordinates of more than 1,000 stars. Pliny
did not give a date for the appearance of the new star, but the
date of Hipparchus' catalogue is known to be close to 130 BC
(cf Dicks, 1960). As Pliny alludes to the motion of the star,
there is a distinct likelihood that it was a comet. The report
is of little interest to us here since no positional details are
given, and in any case Pliny wrote two centuries after the event;
he may merely be quoting hearsay. For a full discussion see
Fotheringham (1919). Stothers (1976) has recently discussed
possible references to new stars in Roman history. He is of
the opinion that there are two allusions to a supernova in
AD 185, but believes that evidence for other novae and supernovae is flimsy. We feel that even the references to the
supernova of AD 185 are questionable, and in any case they give
us no useful details.

The Late Babylonian astronomical texts, which cover the period
from about 700 to 50 BC, contain large numbers of records of
lunar and solar eclipses, occultations, etc. The principal
collection of these tablets is in the British Museum. Plate 1
shows a small fragment which was originally part of an astronomical diary for the year 175 Seleucid (137 - 136 BC). On lines
13 to 15 can be found the most accurate description of a solar
eclipse from ancient or medieval times (for a full translation
see caption to Plate). We learn from Professor Abraham Sachs
of Brown University, USA, who is a leading expert on the Late
Babylonian texts, that there are a few observations of comets
among the Babylonian inscriptions, but the texts which have been
studied contain no references to novae or supernovae. This is
possibly a result of the fragmentary condition of the tablets
because bright novae, as well as supernovae, are rather rare
events (see Chapter 3).

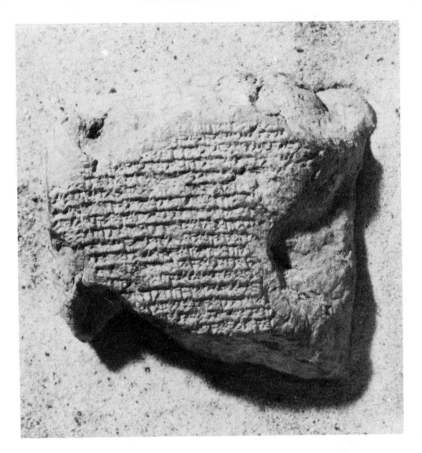

Plate 1.   A Babylonian tablet referring to a
solar eclipse in 136 BC.   Lines 13 to 15 may
be translated as follows (Sachs, personal
communication) :
"Daytime of the 28th the north wind blew.
Daytime of the 29th, 24 uš after sunrise, a
solar eclipse beginning on the south west side
.... Venus, Mercury and the Normal Stars
(i.e. the stars which were above the horizon)
were visible; Jupiter and Mars, which were in
their period of disappearance (i.e. between
last and first visibility) were visible in
that eclipse .... (the shadow) moved from
south west to north east. (Time interval of)
35 uš for obscuration and clearing up (of the
eclipse). In that eclipse north wind which
....."
The unit of time (uš) is the interval of time
required for the heavens to turn through 1°.

Recently Brandt et al. (1976) have drawn attention to various
cave paintings by American Indians which could possibly represent the supernova of AD 1054. Although this is an attractive
hypothesis, the pictures tell us nothing about the star (see
Chapter 8).

The first of the viable sources of historical records of new
stars which we shall consider are the Medieval European monastic
chronicles.

From about AD 1000 there was a flowering of literary activity in
the numerous monasteries scattered throughout Europe; many
monasteries began to keep chronicles of local events. These
reports are often very detailed and make fascinating reading.
Many of the earlier annals were published more than a century
ago in the compilations Monumenta Germaniae Historica (Pertz,
1826→) and Rerum Italicarum Scriptores (Muratori,1723→), both of
which are well known to students of medieval European history.
On the other hand, the numerous later chronicles commencing
after about AD 1200 are relatively inaccessible.

The chronicles are in the main concerned with affairs of the
monastery and local events. However, occasional reference is
made to earthquakes and the more remarkable celestial phenomena,
especially large solar eclipses and bright comets. Frequently
an eclipse or comet is regarded as the major event of the year.
(For translations of many of these records see Newton, 1972, and
Muller and Stephenson, 1975). Occasionally annalists show a
considerable interest in astronomical matters, recording aurorae,
meteors and even occultations, as well as the more striking
phenomena, but such curiosity is rare. As far as the appearance
of a new star is concerned, scarcely less concern could have
been shown. Either the chroniclers did not have adequate
knowledge of the constellations to recognise a new star-like
object, or they held such matters in little regard. To give an
example of the former attitude, Simeon of Durham, a 12th century
English chronicler, describes an occultation of a certain
"bright star" by the totally eclipsed Moon. This occurred some
four centuries before his time - the recorded date corresponds
to AD 775 November 24 - but although the source of Simeon's
account is unknown, the description is clearly that of an eyewitness (Arnold, 1885). Computation by the authors shows that
the "bright star" was in fact the planet Jupiter, but it is
evident that the observer was ignorant of its planetary nature.
We might imagine that had a new star suddenly appeared in the
sky, an observer of this calibre would not have recognised the
fact, unless the star was extremely brilliant.

A particularly intriguing work is the continuation of the
chronicle of Cosmas of Prague by a canon of the cathedral of
Vysehrad in the city. This continuation, which covers the period
AD 1126 to 1142, is published in volume 9 of Monumenta Germaniae
Historica. Our author, who obviously had a keen interest in
astronomy, and is writing as an eyewitness, records observations
of solar and lunar eclipses, aurorae, meteors and parhelia

(mock Suns). However, it is evident from his writings that he
has no more than a hazy knowledge of the planetary motions,
and cannot even identify the planet Venus with certainty. In
the year AD 1131, the canon reports two strange stars, but gives
only vague locations. He writes, "I do not believe that there
are men who know which among them was called Lucifer" (i.e.
Venus). In fact, it seems probable from the directions in which
the "stars" appeared that each was Venus.

Before we criticise, we might well ask ourselves how many
educated people today having no particular interest in astronomy
can point out Jupiter or even Venus. Many will remember the
close conjunction of these two planets which took place early
in 1975. On the evening of February 17, the minimum separation
was less than 0.2 deg. and the spectacle aroused considerable
interest. However, it seems likely that even at the present
time a solitary new star, unless it was unusually bright, would
pass unnoticed to all but the persistent skywatcher.

There is a further, less obvious explanation why we find
scarcely any references to new stars in the chronicles of medieval Europe. We have ample evidence, mainly from the Far East,
that two bright supernovae were visible in the 11th century AD.
The star of AD 1054 appeared in Taurus, and was thus well placed
for northern observers. When it first appeared it would be
prominent in the east before dawn. There are no known European
records of it (see Chapter 8), possibly because it does not seem
to have been much brighter than Venus. On the other hand, the
star of AD 1006 was so brilliant that it could not have failed
to attract widespread attention. It appeared in the southern
constellation of Lupus (declination -37°), but despite its low
altitude, it would be very prominent south of about latitude
48°N and even a casual observer could scarcely have failed to
be impressed by it (see Chapter 7). Yet to our knowledge the
only certain European records of the star are to be found in
the chronicles of the monasteries of Benevento (41°N) and
St. Gallen (47°N).

It seems that the Aristotelian concept of a perfect, changeless
celestial vault was firmly rooted throughout Christendom until
the Renaissance. Sarton, in a passage quoted by Needham (1959),
the source of which is not specified, is particularly good on
this. He writes, "The failure of medieval Europeans and Arabs
to recognise such phenomena (i.e. new stars) was not due to any
difficulty in seeing them, but to predjudice and spiritual
inertia connected with the groundless belief in celestial per-
fection".

Whatever the explanation, the paucity of medieval European
records of novae and supernovae is disappointing in the extreme,
for the number of eclipse and cometary observations is enormous.
Probably the best list of new stars in European chronicles is
that of Newton (1972), which is compiled largely from Monumenta
Germaniae Historica. However, the observations which he lists
are of little scientific interest and reinforce what we have
already said about the poor state of astronomical knowledge which

prevailed in Europe at the time.

What seems at first sight a reasonably promising observation is
recorded by the annalist Albertus in his chronicle of the monastery of Stade (Germany). This work is published in volume 16
of <u>Monumenta Germaniae Historica</u>. Albertus states that in the
year AD 1245 (he is writing as a contemporary) a star appeared
about May 4. He tells us that it was toward the south in
Capricorn. It was large and bright, but red. It could not be
Jupiter because Jupiter was then in Virgo. Many claimed that
it was Mars because of its colour. After July 25 it was no longer bright and it continued to diminish day by day.

We have studied the observations with the aid of Tuckerman's
(1964) planetary tables. It is evident that the "star" was,
in fact, Mars for the planet was almost stationary in Capricorn
from about May until September and was in opposition on July 21.
Newton (1972) thought that the star was a nova and cited the
gradual loss of brightness as evidence of this, although he
emphasised the need for a check on the planetary positions
before a definite judgement was reached. However, the fading
was due to Mars receding from the Earth after opposition,
although the indicated date seems a little early for noticeable
loss of brightness.

To summarise, European records of possible new stars are of
little value, and, with the exception of the observations made
in AD 1006, are of historical interest only. Further research
into the many unpublished chronicles seems likely to repay few
dividends.

We now proceed to the Arabic writings.

The Arabs were certainly restricted in their outlook by the
influence of Aristotle, but it was by no means a stranglehold,
as in Europe. Medieval Arabian astronomy was far superior to
its European counterpart. As the result of an extensive search
through Arabic chronicles, treatises on astrology, etc.,
Goldstein (1965) was able to bring to light several detailed
accounts of the new star of AD 1006 (see Chapter 7). We learn
from Goldstein (personal communication) that there do not
appear to be any extant Arabic records of the star of AD 1054.
In the hope of making further progress, we have initiated a
search of the literature of the period, but at the present time
we can only express the hope that some useful results may emerge.

Before proceeding to the Far Eastern observations, which prove to
be by far the most valuable, we shall briefly mention the Post
Renaissance European observations.

As far as we are aware, no new star (i.e. nova or supernova)
was recorded in Europe between AD 1006 and 1572. Two supernovae
appeared shortly before Galileo first used a telescope in AD 1609
to study the heavens (AD 1572 and 1604), but no similar event
has been observed in our own galaxy since then. It is with the
advent of the telescope that our survey stops.

When the supernova of AD 1006 was seen, European astronomy was
far behind that of the Orient. However, by the late 16th century
the study of the stars had made great progress in Europe,
whereas in the Far East it had remained virtually static. What
can be described as truly scientific observations were made by
Tycho Brahe and others in AD 1572, and these enable the position
of the star to be fixed with some precision. Further, a remarkably accurate light curve can be drawn from the comparisons
which Tycho made between the brightness of the supernova and
that of other stars and planets (see Chapter 10). Kepler,
Fabricius and others did much the same for the supernova of
AD 1604 (see Chapter 11). As a result we know far more about
these two stars than any which had previously appeared.

Both of these comparatively recent supernovae were observed in
China and Korea (not, it seems, in Japan), but the reports from
these countries are fairly unexceptional apart from the Korean
observations of the AD 1604 event. However, whereas astronomy
in Europe had just blossomed forth, the orientals had maintained
a constant study of the heavens from earliest times. It is no
coincidence that practically all of the new stars recorded in
history were observed only in the Far East.

Because of the fundamental importance of the oriental records in
the study of historical supernovae, it seems appropriate to give
a brief outline of the history of the three countries China,
Japan and Korea.

Oriental history began with China some time around 1500 BC.
Prior to this time it is impossible to distinguish history from
legend. The earliest historical dynasty is the Shang. According to orthodox Chinese history this is thought to have followed
the Hsia Dynasty in 1766 BC, but although there are ample
remains from the Shang, nothing has survived from the Hsia.
Indeed, there is not a single reference to substantiate the
existence of the Hsia Dynasty on the oracle bone fragments from
the Shang.

Perhaps a satisfactory date for the beginning of the Shang
Dynasty would be about 1500 BC. The people of Shang practised
divination on a large scale using tortoise shells and the bones
of various animals. After use, the bones were buried to prevent
defilement. Towards the end of last century, vast quantities of
these "dragon bones", as they were called, bearing a very primitive form of ideographic writing, were discovered near An-yang
at the site of one of the main Shang capitals (Yin). As a
result, we now have a very detailed knowledge of Shang divination, but still only a very incomplete picture of the history
of the period. Because of their nature, these oracle texts
contain very few astronomical records and these are mainly of
eclipses (see Needham, 1959). Needham drew attention to two
oracle bones which seemingly referred to the appearance of a new
star near Antares and its subsequent fading (see Fig. 2.1).
A translation of the illustrated text is given by Needham as
follows :

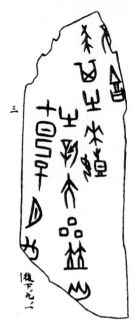

Fig. 2.1. An oracle bone, seemingly referring
to the appearance of a new star.

"On the 7th day of the month, a chi-szŭ day, a great new star
appeared in company with Antares". Antares is described here as
the "Fire (Star)", a name which is common in very much later
texts. However, Jao Tsung-i (1955) suggests that this is not an
astronomical observation, but a question to the ancestors about
making offerings to the "Great Star" and to the "Fire (Star)".

The precise interpretation depends on how we read the character
hsin, which could either mean "new" or "sacrifice". If the
latter is correct, the "Great Star" could possibly be Venus or
Sirius (we cannot be sure). Even if the former alternative is
preferred, the uncertain duration makes the star scarcely worth
considering as a supernova candidate (see Chapter 3). The
first reliable observation of a new star was not made until more
than a thousand years later.

The Shang Dynasty gave way to the Chou Dynasty about 1100 BC.
This commenced almost immediately to establish a feudal system,
which was the great strength of the dynasty in the first few
centuries. However, after about 500 BC the individual feudal
states became fully independent and the period known as that
of the "Warring States" was ushered in. It was not until
221 BC that the empire was finally unified by Ch'in Shih-huang
who became the first emperor of China. From this time onwards
China has seen several partitions, but by and large the country

has been a united empire under the rule of one man.

A list of the various Chinese dynasties is given in Table 2.1.
In this table, the exact year given for the beginning of a
dynasty may differ slightly from that given elsewhere. Much
depends on the precise historical circumstances. Thus the
Ch'in Dynasty was brought to an end in 206 BC by the adventurer
Liu-pang, but, because of rivalry, it was not until 4 years
later that he was proclaimed emperor (of the Han Dynasty).
Again, the Yüan or Mongol Dynasty might be said to begin with
the reign of Genghis Khan (AD 1206), but only in AD 1279 did the
Mongols become absolute rulers of China when the Sung Dynasty
was finally extinguished. In order to avoid somewhat disconcert-
ing gaps in Table 2.1 we have normally taken the commencement
of a dynasty as immediately following the previous one. The
single exception concerns the Kin Dynasty. In AD 1141, the
northern half of China was lost by the Sungs to the Kin Tartars.
The two dynasties co-existed until AD 1234, when the Mongols
overthrew the Kins.

Detailed history of China does not commence until the beginning
of the (Former) Han Dynasty. The Chou Dynasty seems to have
been extremely civilised, and presumably left numerous written
records, but most of these were destroyed at the direction of
Ch'in Shih-huang in 213 BC. On the advice of his prime minister,
Li Szǔ, the emperor issued an order for the "Burning of the
Books". This was a systematic seizure and burning of state
records and any historical literature which did not concern the
state of Ch'in. The object was to eradicate all memory of the
former states which had vied with the emperor's own state of
Ch'in for supremacy.

Ch'in Shih-huang was successful in destroying feudalism, but
the people would not long tolerate such a totalitarian regime.
Soon after his death in 209 BC, a series of rebellions culminated
in the downfall of the Ch'in Dynasty and the establishing of the
Han Dynasty, which apart from a short interregnum was to endure
for more than four centuries. However, the holocaust which
Ch'in Shih-huang instituted has left irreparable gaps in the
history of ancient China. Presumably many further works perished
in the flames when the Ch'in capital of Hsien-yang was destroyed
in 206 BC. At any rate, little survives of Pre-Han history
apart from the Ch'un-ch'iu (together with its enlargement, the
Tso-chuan) and the Shih-chi. The Ch'un-ch'iu is the annals of
the state of Lu, the birthplace of Confucius (hence its preservation), and covers the period 722 - 481 BC. The Shih-chi
("Historical Record") is a history of China from earliest
mythical times down to 122 BC.

From the beginning of the Former Han Dynasty we possess remarkably detailed historical records. Much is due to Szǔ-ma Ch'ien
(145 - 86 BC), who, in writing his monumental Shih-chi, set
the pattern for all subsequent dynastic histories of China. It
became standard practice for the official history of a dynasty
to be compiled during the succeeding dynasty, usually by the
Bureau of Historiography. Occasionally the compilation of a

Dynasty	Dates
Shang	c.1500 BC – c. 1100 BC
Chou	c.1100 BC – 481 BC
Chan-kuo (Warring States period)	481 BC – 221 BC
Ch'in	221 BC – 206 BC
Ch'ien-han (Former Han)	206 BC – AD 9
Hsin	AD 9 – 23
Hou-han (Later Han)	23 – 220
San-kuo (Three Kingdoms period)	220 – 265
Chin	265 – 420
Liu-sung	420 – 479
Nan-pei-ch'ao (Northern and Southern Dynasties)	479 – 581
Sui	581 – 618
T'ang	618 – 907
Wu-tai (Five Dynasties period)	907 – 960
Sung	960 – 1279
Chin (Tartar Dynasty)	1127 – 1234
Yüan (Mongol Dynasty)	1279 – 1368
Ming	1368 – 1644
Ch'ing (Manchu Dynasty)	1644 – 1911

Table 2.1.   A schematic list of Chinese Dynasties

history might be undertaken centuries after the fall of the
dynasty, but, generally speaking, the compilers had free access
to all existing official records. The early invention of print-
ing firstly using wooden blocks (possibly before AD 700) and
later using movable type (from about AD 1050) has contributed
enormously to the preservation of some 2,000 years of continuous
history.

The 26 dynastic histories, together with their dates of writing
(sometimes approximate) are given in Table 2.2. The source is
Han Yu-shan (1955). A typical dynastic history contains four
major divisions : Pên-chi ("Basic Annals") ; Piao ("Chronological Tables"), Chih ("Treatises"), and Lieh-chuan ("Biographies"). Of these, the Chih mainly concern us here since they
are a fundamental source of astronomical records. The sources of
later Chinese history are numerous, but in the search for
astronomical records we can seldom do better than consult the
dynastic histories. For this reason, in discussing Chinese
historical sources we shall concentrate on the official histories, with a brief mention of the Shih-lu ("Veritable Records")
which are available from some of the later dynasties.

Considering the four major divisions of a dynastic history, the
Pên-chi are essentially annals of the emperors and are concerned
with important affairs of state. Events are listed in chronological order, reign by reign. The term Piao, which may be translated literally as "to make manifest", is usually applied to
the chronological tables. The name is taken from the object of
these tables - to make manifest the essentials of the history.

The subjects of the Chih are diverse : rites, music, the calendar, astronomy, geography, law and punishments, the civil
service - to name but a few. Frequently the Chih commences
with the astronomical treatise, emphasising its fundamental
importance, although this is by no means always the case.

By far the largest division of a history is the Lieh-chuan.
The individual biographies are classified, e.g. scholars,
officials, eminent women, artists, technicians. Such great
emphasis on biography underlines the Chinese attitude to history.
As Han Yu-shan (1955) puts it, "The worth of individuals and
human achievements is considered crucial in the making of histories".

A word of caution : The dynastic histories present history in
a unique and readily accessible form. However, the Chinese
conception of history has its drawbacks. In reading a dynastic
history one is always aware that what is written is what the
official historian saw fit to preserve. In particular, it is
clear that many of the astronomical records are mere summaries
of the original observations with considerable loss of detail.
However, we must not be over critical. The present authors have
found the Chinese dynastic histories a seemingly inexhaustible
goldmine of astronomical observations, without equal anywhere in
the world.

History	Period	Written	
1 Shih-chi	(Down to 122 BC)	104 – 87 BC	+
2 Han-shu	Former Han Dynasty	AD 58 – 76	+
3 Hou-han-shu	Later Han Dynasty	398 – 445	
4 San-kuo-chih	Three Kingdoms period	285 – 297	
5 Chin-shu	Chin Dynasty	644	*
6 Sung-shu	Liu-sung Dynasty	492 – 493	+
7 Nan-ch'i-shu		489 – 537	
8 Liang-shu	Northern	628 – 635	+
9 Ch'en-shu		622 – 629	+
10 Wei-shu	and	551 – 554	+
11 Pei-ch'i-shu	Southern	627 – 636	+
12 Chou-shu		c. 630	+
13 Nan-shih	Dynasties	630 – 650	
14 Pei-shih		630 – 650	
15 Sui-shu	Sui Dynasty	629 – 636	*
16 Chiu-t'ang-shu	T'ang Dynasty	940 – 945	*
17 Hsin-t'ang-shu	"	1043 – 1060	
18 Chiu-wu-tai-shih	Five Dynasties period	973 – 974	*
19 Hsin-wu-tai-shih	"	1044 – 1060	
20 Sung-shih	Sung Dynasty	1343 – 1345	*
21 Liao-shih	Liao Dynasty	1343 – 1344	*
22 Kin-shih	Kin Dynasty	1343 – 1344	*
23 Yüan-shih	Yuan Dynasty	c. 1370	*
24 Hsin-yüan-shih	"	1890 – 1920	
25 Ming-shih	Ming Dynasty	1672 – 1755	*
26 Ch'ing-shih-kao	Ch'ing Dynasty	1914 – 1927	*

Table 2.2.   The twenty-six Chinese Dynastic Histories

* Compiled by the Bureau

+ Compiled by Official Decree

We have mentioned in passing the Shih-lu or "Veritable Records".
These were written up at the end of each reign and were based
on the Ch'i-chü-chu ("Diaries of Activity and Repose" of the
emperor). The Ming-shih-lu and the Ch'ing-shih-lu still exist
in their entirety, but these contain very little of astronomical
interest. In this respect they are in marked contrast to their
Korean equivalent, the Yijo Sillok (see below).

It is necessary to explain why astronomical observation was
practised on such a large scale. From earliest times a "Grand
Historian" (Ta-shih) was appointed. According to Han Yu-shan
(1955), the title Ta-shih is to be found among the Shang oracle
bone inscriptions. During the Chou Dynasty, and possibly during
the Shang, the Grand Historian had a double duty to perform.
Since natural phenomena were regarded as closely related to human
affairs, the historian recorded both natural happenings and
human events. In particular, he performed the duties of astronomer and astrologer. During the Former Han Dynasty, by which
time affairs of state had become more complex, the concurrent
duties of the historian were delegated and observation of natural
phenomena was assigned to a separate office, the T'ai-ch'ang
("Office of Sacrificial Worship"). From this time onwards,
the "Astronomical Bureau", as it is usually termed, continued in
operation throughout the rise and fall of dynasties until the
establishment of the modern republic (AD 1912). The officials,
in general, were hereditary, a factor which no doubt made the
system resistant to change.

One of the functions of the Astronomical Bureau was the main-
tenance of the calendar. No doubt this was the main reason for
making the numerous eclipse observations which are recorded
in the various astronomical treatises. However, the astrological
interpretations which accompany many of the eclipse records,
and the vast number of observations of no calendar importance
(e.g. occultations, planetary conjunctions, sunspots, comets,
novae, aurorae) for which astrological prognostications are
also supplied, make it clear that divination was the principal
object of the Bureau. Ho Peng Yoke (1969) has made a detailed
study of the Astronomical Bureau in Ming China. Commenting on an
account in the Ming-shih, he remarks, "With the help of the
Bureau Staff they (i.e. the Director and Deputy Director of the
Bureau) observed or took measurements on the Sun, the Moon, the
stars and asterisms, winds and clouds, and the colour of
vapours, and submitted confidential reports to the Emperors
whenever there were abnormalities". Needham (1959) quotes a
fascinating account of the workings of the Ch'ing astronomical
bureau as the Jesuit Lecomte found it in AD 1696:
"They still continue their observations. Five Mathematicians
spend every night on the Tower in watching what passes overhead;
one is gazing towards the Zenith, another to the East, a third to
the West, the fourth turns his eyes Southwards, and a fifth
Northwards, that nothing of what happens in the four Corners of
the World may scape their diligent Observation. They take notice
of the Winds, the Rain, the Air, of unusual Phenomenas, such as
are Eclipses, the Conjunction or Opposition of Planets, Fires,
Meteors, and all that may be useful. This they keep a strict

Accompt of, which they bring in every Morning to the Surveyor of
the Mathematicks, to be registered in his Office".

In essence, this state of affairs had remained virtually
unchanged since at least Han times. Because of its importance,
the Imperial Observatory was normally built adjoining the Palace
(see Fig. 2.2). As the capital was changed, (usually at the
beginning of a new dynasty), a new observatory would be built.
There is good reason to believe that practically all of the
astronomical observations recorded in the various astronomical
treatises were made at the capital. The workings of the
Astronomical Bureau seem to have been a closely guarded secret
(cf. Needham, Wang-ling and Price, 1960). Because of the pre-
eminent position of the Bureau, falsification of astronomical
observations for astrological purposes was possible, and indeed
probable. How else can we explain the following list of
"observations" recorded in a Korean work, the T'aejŏng Sillok
("Chronicle of the Reign of King T'aejŏng "). The year
corresponds to AD 1406. (We have converted all dates to the
Julian Calendar.)

Fig. 2.2. Plan of the T'ang capital of Lo-yang.

August 10.   "The Moon invaded Mars".
August 14.   "The Moon invaded Mars".
August 17.   "The Moon invaded Mars;   Jupiter and Mars
              invaded one another".
August 21.   "Mars invaded Jupiter".
August 25.   "The Moon invaded Jupiter".

Computation shows that only the last two dates are correct. It
seems that the King and court officials had virtually no knowledge of astronomy. This is by far the most glaring example of
fabrication that we are aware of, but others have been noted (cf.
Bielenstein, 1950).

The astronomical treatises in the dynastic histories show that
almost every conceivable kind of observation that could be made
with the unaided eye was reported by the Bureau. Most treatises
adopt a detailed classification of observations. Thus the
Chin-shu treatise contains 9 sub-sections : "thunder-like
noises"; eclipses of the Sun; solar haloes and sunspots; lunar
changes (eclipses, haloes, etc.); planetary conjunctions; daylight appearances and conjunctions of the Moon and five planets
with asterisms; "Ominous stars" and "guest stars" (comets and
novae); meteors; and "extraordinary clouds and vapours" (usually
aurorae).

There are frequent references to the more spectacular astronomical phenomena (especially eclipses) in the Pên-chi, or Imperial
Annals, but these are couched in layman's terms and are practically devoid of descriptive details. Occasionally biographies
(e.g. of astronomers) mention astronomical events, but this is
rare. The astronomical treatises prove to be our most profitable
source.

The first astronomical records from Japan date from the 7th
century AD. At this time there was no permanent capital, but on
the death of each emperor, the capital was transferred to a new
site. The various capitals were all in the Nara Plain, near
Kyōto. By AD 710, the first regular capital was established at
Nara. This was modelled on the residence of the T'ang emperor
at Ch'ang-an. In the preceding centuries Chinese and Korean
cultural influences had been profound. Particularly important,
from about AD 400 onwards the Chinese system of writing had
been introduced by way of Korea. Practically all of Japanese
astronomical records which we have studied are written in
Classical Chinese.

By the end of the 8th century (AD 794), the capital was moved
to Heian (later known as Kyōto). This was again modelled on
the Chinese capital of Ch'ang-an, and remained the residence of
the Japanese emperors until AD 1869. Towards the end of the
12th century, military governors took control of the country.
While the emperor continued to live at Kyōto, the seat of the
Shogun, or "generalissimo" was Kamakura, near the site of
present-day Tōkyō.

There is no history of Japan comparable with the dynastic histories of China. The astronomical records are in general scattered in a variety of works, such as diaries of courtiers and priv-
ately written histories. A valuable source is the Dainihonshi
("History of Great Japan"), compiled in AD 1715. This contains
a section devoted to astronomical records. However, the study
of Japanese astronomical observations owes much to the patient
research of Kanda Shigeru, who published an exhaustive compilation in 1935. Kanda made a thorough search through histories,
diaries, temple records, etc., and it would be very difficult
to improve on his work. In the present book we have used Kanda
as an index to all our sources of Japanese records, although
in almost every case we have consulted the original works which
Kanda cites.

The Chinese system of naming the constellations (see below)
and Chinese astronomical terms were in use in Japan certainly by
the 8th century AD. Generally speaking, Japanese records are
similar in style to those of China. Once more, the motive was
mainly astrological, but prognostications referred to Japan
rather than the country of origin of the system. Particularly
after about AD 800, astronomical records are quite frequent,
but at no time do we find the degree of regularity or observational accuracy achieved in China.

Korean history (as distinct from legend) may be conveniently
subdivided into four distinct periods: the epoch of the Three
Kingdoms; the Silla ascendancy; the Wang Dynasty; and the
Yi Dynasty. Around 50 BC, largely under Chinese influence,
the various tribes of the peninsula amalgamated into three
independent kingdoms - Silla in the south-east, Paekche in the
south-west, and Koguryŏ in the north, bordering with China. It
was probably about this time that the Chinese script was introduced, but it was not until the 4th century AD that this
became widely disseminated. The three states were almost
continually at war with one another and with China and Japan
in the early centuries.

Around AD 700, in alliance with T'ang China, Silla conquered
Paekche and Koguryŏ. However, some surviving Koguryŏ retainers
established a new dynasty called the Parhae near the Chinese
border. Thus the Korean peninsula and Southern Manchuria were
divided between the two powers - Silla in the south and Parhae
in the north. The gradual decline of Silla in the 9th century
led to the establishment of a new kingdom named Koryŏ in AD 918.
Wang Kŏn, its first king (who later took the title of T'aejo),
successfully unified the whole country for the first time in
AD 936. His capital was Sŏngdo (Kaesŏng). The Wang Dynasty was
to endure for more than three centuries- until AD 1392 when
General Yi Sŏng-gye overthrew the regime and established the Yi
Dynasty with its capital at Han-yang (Sŏul). The country then
took the modern name of Chosŏn. The Yi Dynasty lasted until
the Japanese annexation in AD 1910.

The Samguk Sagi ("History of the Three Kingdoms"), which covers
the period until the commencement of the Wang Dynasty was

compiled in AD 1145 by order of King Injŏng of Koryŏ. It
contains a few astronomical records but these are of doubtful
reliability. One solitary astronomical structure exists from
the Three Kingdoms period. This is in Kyŏngju, the former
capital of Silla, and was built in AD 647 during the reign of
Queen Sŏngdok of Silla. The observatory, illustrated in Plate 2,
is called Ch'ŏmsŏng-dae, literally "Tower for Star Observation".
Its purpose has recently been questioned by Kim Yong-woon (1974),
but the interpretation of the structure as an observatory still
seems most likely. If so, it is the oldest existing observatory
in the Orient.

Coming to the history of Koryŏ, the Koryŏ-sa ("History of the
Kingdom of Koryŏ") is an authoritative work patterned on a
Chinese dynastic history. It was compiled in the early decades
of the Yi Dynasty by official decree and was finally completed
in AD 1451 after many revisions. This work, which is by far
the most valuable history of the period, contains an astronomical treatise of three chapters. From this it is clear that
after about AD 1,000 a high level of attainment was reached.
As in Japan, Chinese astronomy was adopted, and indeed the
astronomical treatise of the Koryŏ-sa is indistinguishable from
a similar treatise in the Chinese dynastic history.

The most valuable source of astronomical records during the Yi
dynasty is the Yijo Sillok. Sillok is the equivalent of the
Shih-lu in China, i.e. "Veritable Records" written up at the
end of each reign (see above). However, the Korean work is
very much more detailed than its Chinese counterparts. In
the Sillok, astronomical records are not contained in a separate
section, but are included in chronological order along with
affairs of court and state. Fortunately, summaries of the
various observations are classified in the Chŭngbo Munhŏn Pigo,
a late Korean historical summary, which acts as a useful index
to the records in the Sillok.

It might be pointed out that all four Korean works mentioned are
written in Classical Chinese. The modern Korean alphabet,
known as Han-gŭl, and consisting of 11 vowels and 28 consonants
was invented by King Sejŏng in the 15th century AD. This
achieved considerable popularity, but met with opposition from
scholars, hence its neglect in literary works.

We shall conclude this chapter with discussions on the Chinese
calendar and the Chinese conception of the constellations.

The Chinese calendar, which was adopted from an early period in
both Korea and Japan, is fundamentally luni-solar. However, an
independent system of counting days makes reduction to the
Western calendar a relatively simple operation. In China,
years were originally numbered from the beginning of the reign
of a king (before the Unification in 221 BC) or emperor (after
the Unification). Emperor Wên, of the Former Han Dynasty
(179 - 156 BC) was the first to establish reign periods, i.e.
subdivisions of a reign. There were usually two or three such
periods in a reign, although the exact number could vary from

Plate 2. The "Tower for Star Observations" at
Kyŏngju. Built in AD 647, it is the oldest
existing observatory in the Orient.

only one to more than ten. From the 8th century AD onwards,
Japanese emperors adopted a similar system, but it was never
introduced in Korea. It might be mentioned here that the ruler
of Korea was always known as a king, never an emperor. Tables
of reigns and reign periods are readily available, e.g. Moule
and Yetts (1957) or Tsui Chi (1947) for China and Tchang (1905)
for Japan and Korea.

Probably from as early as the Shang Dynasty in China, alternate
months of 29 and 30 days, with occasional adjustments, enabled
the calendar to keep pace with the Moon (the mean length of the
synodic month - from one new Moon to the next - is 29.5306 days).
Certainly by the Chou Dynasty (if not before), an astronomical
new Moon, rather than an observed crescent, began the month,
and as a result all but a few solar eclipses are recorded on
the first day of the lunar month. Most years contained 12
months. The problem of the incompatibility of the synodic
month and the tropical year (365.2422 days) was surmounted by
the insertion of an intercalary month every $2\frac{1}{2}$ years or so.
From Han times, intercalation was very regular the object being
to make the new year (beginning of the first month) fall on the
second new Moon after the winter solstice.

Side by side with the luni-solar calendar, an independent system
of day numbers is in operation. This 60 - day cycle, similar
in many respects to our week, was in use during Shang times.
The cycle was formed by the combination of a series of 10 symbols
- the "celestial stems" (t'ien-kan) - with another series of
12 symbols - the "terrestrial branches" (ti-chih). By taking
the symbols in pairs (one from each series) in cyclical order,
60 possible permutations are obtained.

The 10 t'ien-kan are as follows : chia, i, ping, ting, wu, chi,
kêng, hsin, jen, kuei. The 12 ti-chih are :tzǔ, ch'ou, yin, mao,
ch'en, szǔ, wu, wei, shen, yu hsü, hai. We have shown the
full sexagesimal cycle in Table 2.3. To give an example of the
use of this table, the day kêng-ch'en is the 17th day of the
cycle. It should be noted that many combinations cannot occur,
e.g. there is no day jen-wei.

Analysis of the eclipse records in the Ch'un-ch'iu or "Spring
and Autumn Annals" (722 BC - 480 BC) demonstrates that the
cycle has been unbroken since this period (see Muller and
Stephenson, 1975). There is really no reason for suspecting
that any interruption occurred during the Shang or early
Chou dynasties, so that it seems probable that the cycle has
had a continuous existence for more than three millennia.
So important has the sexagesimal cycle proved that in oriental
records the day of the month, unless it happens to be the first
or last day is seldom given (except in Japan), whereas the
cyclical day is almost always specified. The same system was
in use for numbering years, but it never achieved the popularity
of the day-numbering system. It is interesting to note that the
12 ti-chih are used for naming the double hours. The first
double hour (tzǔ)is the period between 11 p.m. and 1.a.m. local
time, and so on throughout the day. Thus wu corresponds to

	tzu	ch'ou	yin	mao	ch'en	szu	wu	wei	shen	yu	hsu	hai
chia	1		51		41		31		21		11	
i		2		52		42		32		22		12
ping	13		3		53		43		33		23	
ting		14		4		54		44		34		24
wu	25		15		5		55		45		35	
chi		26		16		6		56		46		36
keng	37		27		17		7		57		47	
hsin		38		28		18		8		58		48
jen	49		39		29		19		9		59	
kuei		50		40		30		20		10		60

Table 2.3. Days of the Sexagesimal Cycle

9 a.m. to 11 a.m.

Tables produced by Hsüeh Chung-san and Ou-yang I (1956) enable
the rapid conversion of Chinese dates to the Western calendar.
(Julian or Gregorian) for any date between AD 1 and 2000.   These
tables are equally useful for converting Japanese or Korean
dates when used in conjunction with a list of reigns or reign
periods.   Numerous examples of dates expressed in the oriental
style are given in succeeding chapters.

Some time before the Han Dynasty (just when is impossible to say
owing to the paucity of extant records), an amazingly complex
astronomical / astrological system evolved in China.   The sky was
subdivided into several hundred small asterisms, each containing
an average of 5 or 6 stars.   Around AD 275 the Astronomer Royal
estimated from early star maps then in existence that there were
283 star groups and 1464 stars (Ho Peng Yoke, 1966).   Later
estimates vary considerably, but these figures are fairly representative.

The names of the various asterisms span the whole spectrum of
life in China, e.g. Ti-wang ("Emperor"), T'ai-tzŭ ("Crown
Prince"), Shang-shu ("Secretaries"), Huan-chê ("Court Eunuchs"),
T'ien-chuang ("Celestial Bed"), Ch'uan-shê ("Guest Houses"),
Nei-ch'u ("Inner Kitchen"), Sha-hsing ("Execution Star"),
T'ien-chiang ("Celestial River") to name but a few, chosen more
or less at random.   A complete list is given by Ho Peng Yoke
(1966).

Celestial events, such as a planet, comet or new star entering a
particular asterism, or the occurrence of an eclipse, were regarded as precursors of terrestrial events involving the corresponding personage, object, state, etc.   The astronomical treatises of the earlier dynastic histories (particularly those of
the Han and Chin) contain detailed prognostications for almost
every observation.   We have selected the following examples from
the Hou-han-shu and the Chin-shu, referring to the solar eclipse
of AD 65 and the new star of AD 369 respectively :
(AD 65).   "8th year (of the Yung-p'ing reign period), 10th month,
(day) wu-wu, the last day of the month.   There was an eclipse of
the Sun and it was total.   It was 11 degrees in (Nan-) tou.
(Nan-) tou represents the state of Wu.   Kuang-ling, as far as
the constellation are concerned, belongs to Wu.   Two years later
Ching, King of Kuang-ling, was accused of plotting rebellion and
committed suicide".
(AD 369).   "4th year of the T'ai-ho reign period of Hai-hsi, 2nd
month.   A guest star was seen at the western wall of Tzŭ-wei.
When we come to the 7th month it finally disappeared.   The interpretation when a guest star guards Tzŭ-wei is assassination of
the Emperor by his subjects.   In the 6th month Huan-wei dethroned
the Emperor who became the "Duke of Hui-hsi" .

It can be seen from these two examples that the delay between
the occurrence of an omen and its acknowledged fulfilment was
usually comfortably long.   Two years was fairly typical,
evidently to ensure the maximum chance of success.

According to Ho Peng Yoke (1966), the belief in astral influence
on state events germinated in the Ch'un-ch'iu period (722 -
480 BC), and became firmly implanted through the teaching of
Liu-hsiang (77 to 6 BC), a well known astronomer and mathematician. In any event, it was to endure for more than two
millennia, and established itself in both Korea and Japan.
Certain emperors were opposed to astrology and divination,
e.g. Wu-ti, who in AD 268 passed an edict to ban the study of
these subjects,and Yang-ti, who about AD 610 ordered all books
dealing with these subjects to be burnt. However, and fortunately for modern science, these enlightened rulers were in the min-
ority. Ironically enough, the astronomical treatises in the
Sung-shu and Chin-shu regard an observation of Mars in AD 287
as an omen of Wu-ti's death; evidently his influence was short-
lived!

The prognostications accompanying the observations seem trivial
now, but it is a direct result of the importance attached to
astrology that we possess such a vast number of observations of
all kinds, and in particular such an impressive list of new star
sightings. Further, since the substance of a particular prognostication depended very much on which asterism the new star
appeared in, Far Eastern records are usually careful to give a
fairly accurate location. Normally only the asterism is named,
which might mean that the star appeared anywhere within an area
of say 100 square degrees of sky, but occasional positions were
measured to the nearest degree by the Korean astronomers (see
Chapter 11). Needham (1959) believes that from the Han onwards
measurements of position were made in degrees, but the histor-
iographers simplified the information they received in writing
the official histories. This may well be the case.

Oriental positions are often expressed in terms of the 28 hsiu
or "lunar mansions". Each lunar mansion covers a range of right
ascensions, obviously averaging about 13 deg. in extent, but the
individual extensions are very variable. The hsiu are named
after asterisms near the celestial equator, and the boundary of
each hsiu is fixed by the location of a certain "determinant
star" of the asterism. The lunar mansions are in order (with
the determinant stars in parentheses) :

1.	Chüeh	(α Vir.)		15.	K'uei	(η And.)
2.	K'ang	(κ Vir.)		16.	Lou	(β Ari.)
3.	Ti	(α^2 Lib.)		17.	Wei	(41 Ari.)
4.	Fang	(π Sco.)		18.	Mao	(η Tau.)
5.	Hsin	(σ Sco.)		19.	Pi	(ϵ Tau.)
6.	Wei	(μ' Sco.)		20.	Tsui-hsi	(λ Ori.)
7.	Chi	(γ Sgr.)		21.	Shen	(ζ Ori.)
8.	Nan-tou	(φ Sgr.)		22.	Tung-ching	(μ Gem.)
9.	Niu	(β Cap.)		23.	Yü-kuei	(θ Cnc.)
10.	Hsü-nü	(ϵ Aqr.)		24.	Liu	(δ Hya.)
11.	Hsü	(β Aqr.)		25.	Ch'i-hsing	(α Hya.)
12.	Wei	(α Aqr.)		26.	Chang	(ϵ' Hya.)
13.	Ying-shih	(α Peg.)		27.	I	(α Crt.)
14.	Tung-pi	(γ Peg.)		28.	Chen	(γ Crv.)

The asterisms listed above lie roughly on a great circle. At
the present time, the declinations vary between about -30 and
+30 deg., but around 2500 BC, the circle encompassing the lunar
mansions corresponded reasonably closely to the celestial
equator. There have been attempts to use this method in tracing
the origin of the lunar mansions, but in the absence of confirm-
atory historical evidence, we can no more than hypothesise (for
a discussion see Needham, 1959).

Lunar mansions are frequently used to specify the positions of
new stars. Fortunately, in such instances oriental records are
quite clear to distinguish whether the actual asterism is referred to or the range of right ascensions bearing its name.
When the name of the asterism alone is used, the star group
itself is intended. However, when the character hsiu (i.e.
"lunar mansion") is appended, or a star is described as so many
degrees (tu) within the asterism, the right ascension is implied.

An excellent representation of the oriental asterisms is shown in
Plates 3 and 4. These maps were drawn by Jesuit astronomers in
China in AD 1747 (the charts were originally published by Rigge,
1915). The two hemispheres are shown separately in polar
projection. On the maps, the radial lines represent the boundaries of the 28 lunar mansions. The inner circle on the Northern Hemisphere chart depicts the boundary of the region of perpetual visibility from about 40° N (the latitude of Peking). The
corresponding circle on the Southern Hemisphere chart is the edge
of the south polar region, invisible from North China. Evidently
the asterisms within this latter region were drawn from Western
observations made in the Southern Hemisphere. It will be noticed that no attempt is made to distinguish between stars of
different magnitude. This feature is characteristic of practically all oriental star maps.

In our discussion of the new star observations we have made use
of a variety of oriental star maps in addition to the above.
A list of these is as follows :
1.   An early Jesuit map (date around AD 1600) preserved in the
Consiglio Nazionale delle Ricerche at Bologna, Italy. Our
attention was first drawn to this by Professor G. Palumbo. The
chart is similar in form to the later Jesuit map discussed above.
2.   A series of maps in the Ku-chin-t'u-shu-ch'i-cheng, a
Chinese encyclopedia dating from about AD 1725, but containing
pre-Jesuit star maps from more than a century earlier. These
maps are crudely drawn, but they are particularly valuable since
they distinguish three different types of star "magnitude" -
an almost unique feature.
3.   A similar series of late pre-Jesuit charts in another
encyclopedia, the San Tsai Tu Hwei. Here no attempt is made to
represent star brightness.
4.   A Korean planisphere of AD 1395. This is an exact copy of
a very early Chinese star map. The original stone engraving was
presented to the King of Koguryŏ by the Emperor of China at an
early period. This stone was lost in a river when Koguryŏ fell
to Silla in AD 672. However, a new engraving based on a preserved rubbing of the original, was made in AD 1395. This, and

NORTHERN HEMISPHERE.

Plate 3. A Chinese star chart, drawn by Jesuits in Peking in the 18th century, shows asterisms in the northern celestial hemisphere.

SOUTHERN HEMISPHERE.

Plate 4. A companion star chart to Plate 3,
showing the southern celestial hemisphere.

Plate 5. Bronze planisphere found some time
last century in a Japanese junk. This shows the
asterisms visible from latitude 35° N. The
planisphere, which is about 35 cm. (14 in) in
diameter, is an almost exact copy of a Korean
star map engraved on stone in AD 1395 and still
surviving. This was itself copied from a very
early Chinese map. The position of the equinox
corresponds to about 100 BC, so that the original map is of very ancient origin. The circles
represent the celestial equator, ecliptic and
'Purple Palace' (circle of perpetual visibility).
Other features are the boundaries of the 28
'lunar mansions', represented by radial lines,
and the Milky Way, shown as a curving irregular
band.

a later copy (AD 1668) still survive in a Sŏul museum. Rufus
(1913), who made a detailed study of the planisphere, estimated
the epoch when the chart was first prepared as close to 100 BC
from the position of the equinox and celestial pole.   The
Royal Scottish Museum, Edinburgh, contains a bronze planisphere
which was found in the 1870's in a Japanese junk which was
wrecked on an island off the coast of Japan.   Rufus and Chao
(1944) drew attention to the fact that this planisphere is
either a careful copy of the Korean map or shares a common origin
with it.   The Japanese planisphere is illustrated in Plate 5.
Idealization of certain star groups is evident, but the original
Chinese astrography was apparently still a valuable aid to
navigation some two millennia later.
5.   The Soochow chart prepared in AD 1193 and engraved on stone
in AD 1247.   This map is rather idealized and several of the
asterisms appear to be displaced to allow space for their names.
An excellent rubbing is illustrated by Rufus and T'ien Hsing-
chih (1945).
6.   A fairly accurate series of maps contained in the Hsin-i-
hsiang-fa-yao, a work by Su Sung and dating from AD 1092.   These
are probably the earliest printed star maps (cf Needham, 1959).
7.   The crude star maps which form part of a vast collection
of manuscripts which Sir Aurel Stein discovered last century in
Tunhuang (Kansu).   The date of these maps is AD 940 (Needham,
1959).
8.   Hsing-ching ("Star Manual").   This work, which is contained
in the Taoist Canon, is probably a compilation of Pre-Han
and Han catalogues with later additions up to about AD 600.

Having considered the sources of the historical records, we are
now ready to proceed to a discussion of the new star records
themselves.

Chapter 3

A CLASSIFICATION OF THE
FAR-EASTERN NEW STARS

Basically, three kinds of new star were recognised in the Far
East. These are k'o-hsing ("guest stars" or "visiting stars"),
po-hsing ("rayed stars" or "bushy stars") and hui-hsing ("broom
stars" or "sweeping stars"). These will be considered in turn.

K'o-hsing (which will be subsequently abbreviated to k'o)' seems
to have been the general term to describe a new star-like object.
The well known new stars of AD 1006, 1054, 1572 and 1604 were
identified in this way and we might thus expect k'o to be synonomous with novae and supernovae. On the other hand, there are
frequent references to moving k'o throughout oriental history
(more than 20 are catalogued by Ho Peng Yoke, 1962), so that
usage of the term must be treated with caution. The nucleus of
a comet resembles a star, so that if no tail is evident confusion seems possible.

Po-hsing (po) is the standard term to describe an apparently tail-
less comet. In the astronomical treatise of the Chin-shu (the
official history of the Chin Dynasty) we find the following:
    "By definition a comet pointing towards one particular
direction is a Hui Comet and one that sends out its rays evenly
in all directions is a Po comet".

Ho Peng Yoke (1966) attempted to explain a po as a tailed comet
seen almost head-on, i.e. in opposition to the Sun, but this can
be the explanation for only a small proportion of such stars.
The records of po indicate that they could appear at any elongation from the Sun, and furthermore were frequently visible for
several months. In the catalogue of Ho Peng Yoke (1962) the
motions of some 10 po and k'o for three months or more are
described. Reading the various descriptions of these objects,
it is difficult to avoid the conclusion that some comets can
remain visible for an extended period without developing an
appreciable tail. We occasionally find accounts of po which
seem to imply that the star remained stationary for a number of
days - i.e. there is no reference to motion. The possibility
thus exists that some of these stars were novae rather than
comets. However, it should be remembered that most oriental
records which are available to us are mere summaries of the
original observations, possibly with considerable loss of detail.
As with k'o some care must be exercised in classifying po in
modern terms.

Hui-hsing (hui) is the general term for a tailed comet. In the
same section of the Chin-shu as the previous definition we find
the following description of a hui:

"Its body is a sort of star, while the tail resembles a
broom.   Small comets measure several inches in length, but the
larger ones extend across the entire heavens".

Since the use of the term hui implies that the star had a definite tail, the probability of a nova or supernova being described
in this way is obviously very low.   Many records of hui give an
estimate of the length of the tail, frequently several tens of
degrees, and a detailed account of the motion is often given.
On the other hand, often a brief mention of the sighting of a hui,
without any reference to motion, is reported.   Here again we may
be dealing with a brief summary of the original observation.

A particularly remarkable feature of oriental observations of
comets is the extent to which two independent reports of the same
star can vary.   To give a random example, an account of the
motion of Halley's Comet in the Sung-shih (the official history
of the Sung Dynasty in China) is extremely detailed, whereas only
a single sighting is recording in the Koryŏ-sa (the official
history of the Wang Dynasty in Korea).   Numerous similar examples could be quoted.(The year of the above sighting is AD 1066.)

It has been pointed out by Ho Peng Yoke (1962) that the supernova of AD 1572 was called a hui in the Ming-shih-kao, the draft
version of the official history of the Ming Dynasty.   However,
this error was corrected in the history itself (the Ming-shih),
where the star is described as a k'o.

As an alternative to hui-hsing, the term ch'ang-hsing ("long
star") is not infrequently used.   Once again this denotes a
tailed comet, and we occasionally read of ch'ang-hsing with tails
stretching across the sky (for instance, in 135 BC and AD 128).
As for hui, we feel that, except when there seems to be very good
reason to the contrary, all ch'ang-hsing should be regarded as
comets.

Finally, a few miscellaneous "stars", usually nameless, are from
time to time reported in Far Eastern historical sources.   In
such instances it seems best to treat each individual case on its
own merits.

The first catalogue of temporary stars recorded in history,
compiled for its own sake, was probably that of Fujiwara Sadaie,
a Japanese poet-courtier.   Following the appearance of a k'o in
AD 1230 (this was in reality a comet - see below), Fujiwara made
a search of Japanese historical works for references to new stars.
In his diary, the Meigetsuki ("Diary of the Full Moon"), he gave
details for eight such stars ranging in date from AD 642 to 1181.
All are to be found in the entry for AD 1230 December 13.   This
diary is an important source of Japanese records of novae and
supernovae.

In China at about the same time, Ma Tuan-lin in his encyclopedia,
the Wên-hsien-t'ung-kao, included a list of temporary stars in
his section on Chinese astronomical observations.   However, this

list, although very useful to modern astronomers, did not have a
special motive.   In the 18th and 19th centuries a number of
lists of historical new stars were compiled - Pingré (1783),
Biot (1843), Humboldt (1851) and Williams (1871) - however, these
are of little value at the present day.   Perhaps the first
person to realise the significance of temporary stars other than
comets was Lundmark (1921).   He produced a catalogue of 60
suspected novae and supernovae based on earlier secondary sources,
in which he deduced approximate co-ordinates and plotted the
galactic distribution of the stars.   Unfortunately, Lundmark
gives us very little information on his criteria for classifying
stars as novae rather than comets.   His catalogue has been
superseded by modern ones which are based on direct consultation
of original sources.

Kanda Shigeru (1935) made an extensive compilation of Japanese
astronomical observations recorded in a wide variety of sources.
This covers the period up to AD 1600, and includes a list of new
stars (both comets and novae).   Kanda's work forms a valuable
source of references.

The first modern catalogue of possible novae and supernovae is
due to Hsi Tsê-tsung (1955).   This contains ninety entries up to
AD 1690, mainly obtained from Chinese and Japanese historical
materials.   Complete accounts and references are given for each
star, together with estimated co-ordinates, and the galactic
distribution is shown diagrammatically.   Hsi Tsê-tsung
criticised several of Lundmark's selections, but his own catalogue
is far from reliable.   Po and hui are included along with k'o,
almost at random.   Again, certain stars are suggested as supernovae without foundation, while pairs of stars are linked as
possible recurrent novae even though the individual positions
were only approximately recorded.

The Smithsonian translation of Hsi Tsê-tsung's catalogue is
excellent.   However, when using this work it should be pointed
out that po-hsing is translated as "sparkling star".

More recently, in collaboration with Po Shu-jen, Hsi Tsê-tsung
(1965) produced an improved version of his previous catalogue.
Korean sources and a Vietnamese history were also consulted, and
again a total of ninety stars up to AD 1690 were listed.   However, this work suffers from much the same defects as its
predecessor, and like it should be used with caution.   It might
be remarked that the abridged translation by Yang (1966) is far
superior to the, at times, shockingly bad NASA translation.

The brief catalogue of Chu Sun-il (1968) contains summaries of 28
Korean observations of temporary stars up to AD 1665, together
with estimated positions.   However, Chu does not distinguish
between the various types of star which he lists and it is thus
impossible to assess the reliability of any particular sighting
without consulting the original record (or a translation
published elsewhere).

Of the fundamental catalogues, we are left with the two compilations of Ho Peng Yoke (1962, 1970).   Both works are entirely
based on Far Eastern sources, the original giving Chinese,
Japanese and Korean observations up to AD 1600 and the supplement
Chinese observations from AD 1368 to 1911.   The great value of
Ho Peng Yoke's work lies mainly in the impartial nature of his
compilation.   He set out to compile a list of comets and novae,
and it is left to the reader to decide for himself the nature of
any particular object.   This is backed up by an exhaustive
search of the literature and excellent translations.   Further,
the oriental star maps which he provides in his original catalogue, although not always reliable (Stephenson, 1971), enable an
estimate of the uncertainty in the position of any particular
star to be made.

Recently, Pskovskii (1972) has produced a secondary catalogue of
possible novae and supernovae based on the various modern fundamental catalogues discussed above.   Justifiably, Pskovskii
criticises the somewhat naive approach of some of these works,
but his own methods of selection are open to criticism.   For
instance, he includes without justification many po in his
catalogues which were rejected by Hsi Tsê-tsung and Po Shu-jen
(1965).   In particular he states that "no comet has ever been
observed with the unaided eye for more than four months",
implying that all stars of longer duration were novae or supernovae.   In fact, Ho Peng Yoke (1970) gives full details of the
motions of several comets which were observed in the Far East
for longer than four months, up to a maximum of six and a half
months.

The catalogue of pre-telescopic novae and supernovae to be
presented here makes no claim to originality in so far as the
material used is concerned.   It is based on a collation of the
fundamental catalogues discussed above.   However, in view of the
authors' considerable experience in handling oriental astronomical records, it is hoped that the present work is as objective
as possible.   It should be remarked that we have independently
consulted almost all of the original records of the most
important observations.

The only new stars sighted in the West which are included in the
present catalogue are those of AD 1006, 1572, and 1604.   An
Arabic report of the new star of AD 1006, falsely dated as AD 827
by Humboldt (1851) and others, was correctly dated by Goldstein
(1965);   see Chapter 7.   Excluded from the catalogue are the new
stars which are supposed to have appeared in AD 945 and 1246.
Our only authority for these is the 16th century astronomer
Cyprianus Leovitius.   Leovitius did not name his sources, and the
original records have never been uncovered.   Humboldt drew
attention to a new star which a certain Cuspinianus was supposed
to have observed in AD 393.   According to Humboldt, this appeared close to α Aquilae and rivalled Venus.   However, Lynn (1884)
pointed out that Cuspinianus was a 16th century AD writer, and
he traced the original report to the 5th century Chronicon of
Marcellinus.   This states, "A star rising from the north at

cockcrow, and burning like Venus rather than shining, appeared, and on the 23rd day it disappeared." The approximate date is the beginning of September in AD 389. Only a very rough location is given, and Lynn emphasized that "no star near the constellation of Aquila could have risen at cockcrowing" around this time of year. He in fact concluded that the star was identical with a comet described by the contemporary writer Philostorgius, who gives an account of its motion. So much for the reliability of secondary sources!

Coming now to the Far East, the bulk of the new stars which we have catalogued were originally classified as k'o. These include stars of both long and short duration, as well as those of indefinite duration for which there is only a mention of the first sighting. We have listed all k'o for which a celestial position is given and for which there is no definite mention of a tail or motion. It is unfortunate that many reports of k'o give only the general direction (usually one of the four cardinal directions) in which the star appeared and this, of course, is useless. Deciding whether or not a tail is referred to is not always straightforward. Frequently a record may merely state that a star "measured x ch'ih". Actually, ch'ih ("feet") is a linear unit, but Kiang (1972) and Stephenson (1971) have independently shown that, roughly at least, ch'ih and tu (degrees of 365.25 to a circle) were synonomous. To the average eye, a bright star appears to have "rays" of an appreciable extent due to spherical aberration, astigmatism, etc. In order to reduce the number of spurious objects, we have systematically omitted all stars which were said to measure more than 1 ch'ih (roughly twice the Moon's apparent diameter), unless an unusually long duration, with no apparent motion, is reported.

We have excluded all isolated sightings of po. Use of this term implies that the star possessed a definite form, and when no duration is recorded the probability of it being a nova seems low. It is likely that a number of k'o of indefinite duration were comets, but here at least the "correct" term for a nova is applied. Where a po is recorded with a definite duration, but without any hint of motion or tail, there is a reasonable possibility that the star was misidentified, and accordingly we have included such examples in our catalogue.

With two exceptions (hui in 5 BC and AD 247) all hui and ch'ang-hsing are excluded from the catalogue. These two stars remained visible for many days and there is no hint of any motion. Further investigation thus seems desirable. All other such stars for which no motion or tail is mentioned were visible for about 25 days or less (usually much less) and accordingly seem scarcely worth considering as possible novae or supernovae.

It should be pointed out that two well known k'o are omitted from our catalogue. These are the supposed supernovae of AD 902 (Pskovskii, 1963) and AD 1230 (Shklovsky, 1968). The former was considered in some detail by Stephenson (1975). The Chinese record of the star (in the Hsin-t'ang-shu, the "New Book of the

T'ang Dynasty") distinctly states that it changed position -
"During the first month of the second year of the T'ien-fu
reign period, (11 February to 12 March AD 902) a guest star like
a peach was at Tzŭ-wei beneath (i.e. to the north of) Hua-kai.
It gradually moved and reached Yŭ-nu.   On the day ting-mao
(2 March) a meteor (liu-hsing) left Wên-chang and reached the
guest star; the guest star did not move.   On the day chi-szŭ
(4 March) the guest star was at K'ang and guarded it.   In the
following year (ming-nien) it still had not faded away".

The whole record is most difficult to interpret.   We are told
quite specifically that the star moved, so that a tail-less comet
seems to be the obvious interpretation.   Further, all the details
relate to the first month and only at the end do we find that it
was still visible the following year.   This is much too long a
duration for a comet.   We feel that we must read jih ("day") for
nien ("year"), although the characters are very different, so that
the last sentence commences, "On the following day".   All the
events then occurred in the same month and the latter part of the
entry could be merely interpreted as a commentary to the effect
that although the meteor seemed to come in contact with the guest
star it did not disturb it in any way.   On the other hand we may
be dealing with two independent guest stars, but it is impossible
to be definite.   Regretfully we must abandon the AD 902 star as
of uncertain nature.

Ho Peng Yoke (1962) showed quite definitely that the star of AD
1230, which was said to be visible for 4 months (in the Sung-shih,
the official history of the Sung Dynasty) was a comet, by quoting
a parallel record in the Kin-shih (the official history of the
Kin Tartar Dynasty) which gives a detailed description of the
motion of the star.   It should be noted that both histories
describe the star as a hui, the standard term for a comet.   Only
in Japan was it called a k'o.   This example, in which a comet
which was visible for several months, and yet could be recorded
in one history without any mention of motion, suggests that other
similar instances might occur.

Although the AD 902 and AD 1230 new stars are excluded from our
catalogue, all other stars which have at one-time been seriously
proposed as supernova candidates are included.

Coming now to the few references to other types of star, we have
included in our catalogue all stars not in the above categories
for which there is no certain reference to motion or a tail,
regardless of duration.   Here again, it is highly likely that
some comets are included, but this is unavoidable with such
sketchy material.

Generally speaking, it is obvious that the reliability of the
observations included in the catalogue varies enormously.   In
an attempt to allow for this we have devised a somewhat arbitrary
classification scheme.   This will be discussed below.

Table 3.1. Catalogue of pretelescopic galactic novae and supernovae.

REF. NO.	DATE	PLACE	TYPE	DURATION	CLASS	RA (1950.0)	DEC (1950.0)	l	b
01	BC 532 Spring	C	'star'	–	5	20^h 50^m	$-10°$	$40°$	$-30°$
02	204 Aug./Sep.	C	po	10 days	5	14 20	+20	20	+70
03	134 Jun./Jul.	C	k'o	–	4	16 00	-25	350	+20
04	77 Oct./Nov.	C	k'o	–	4	11 10	+75	130	+40
05	76 May /Jun.	C	chu	–	5	1 40	+25	135	-35
06	48 May	C	k'o	–	4	18 40	+25	10	-10
07	47 Jun./Jul.	C	k'o	–	4	4 00	+65	140	+10
08	5 Mar./Apr.	C	hui	70 + days	2	20 20	-15	30	-25
09	AD 61 Sep. 27	C	k'o	70 days	2	14 10	+35	60	+70
10	64 May 3	C	k'o	75 days	2	12 20	-5	290	+55
11	70 Dec./Jan.	C	k'o	48 days	1	9 40	+25	215	+45
12	85 Jun. 1	K	k'o	–	5	–	+65	–	–
13	101 Dec. 30	C	k'o	–	4	9 40	+25	215	+45
14	107 Sep. 13	C	k'o	–	4	6 30	+10	200	0
15	125 Dec./Jan.	C	k'o	–	4	17 10	+10	30	+25
16	126 Mar. 23	C	k'o	–	5	12 00	+10	270	+70
17	185 Dec. 7	C	k'o	20 months	1	14 20	-60	315	0
18	222 Nov. 4	C	k'o	–	4	12 30	0	290	+60
19	247 Jan. 16	C	hui	156 days	2	12 30	-20	295	+40
20	290 Apr./May	C	k'o	–	4	–	+65	–	–

(1)	(2)	(3)	(4)	(5)	(6)	(7)		(8)	(9)	(10)
21	304 Jun./Jul.	C	k'o	-	4	4	20	+15	180	-25
22	329 Aug./Sep.	C	po	23 days	5	12	30	+55	130	+65
23	369 Mar./Apr.	C	k'o	5 months	1		-	+65	-	-
24	386 Apr./May	C	k'o	3 months	1	18	30	-25	10	-10
25	393 Feb./Mar.	C	k'o	8 months	1	17	10	-40	345	0
26	396 Jul./Aug.	C	'star'	50 + days	2	4	00	+20	175	-25
27	402 Nov./Dec.	C	k'o	2 months	2	11	10	+10	240	+60
28	421 Jan./Feb.	C	k'o	-	4	11	30	-15	275	+45
29	437 Jan. 26	C	'star'	-	5	6	40	+20	195	+ 5
30	483 Nov./Dec.	C	k'o	-	5	5	30	0	205	-15
31	537 Jan./Feb.	C	k'o	-	4		-	+65	-	-
32	541 Feb./Mar.	C	k'o	-	4		-	+65	-	-
33	561 Sep. 26	C	k'o	-	4	11	30	-15	275	+45
34	641 Aug. 6	C	po	25 days	5	12	20	+20	265	+80
35	684 Dec./Jan	J	po	2 weeks	5	3	40	+25	165	-25
36	722 Aug. 19	J	k'o	5 days	3	1	00	+60	125	0
37	829 Nov.	C	k'o	-	4	7	50	+15	205	+20
38	837 Apr. 29	C	k'o	22 days	3	7	00	+10	205	+ 5
39	837 May 3	C	k'o	75 days	1	12	10	+ 5	280	+65
40	837 Jun. 26	C	k'o	-	5	18	00	-25	5	0

(1)	(2)	(3)	(4)	(5)	(6)	(7)		(8)	(9)	(10)
41	877 Feb. 11	J	k'o	–	4	23	50	+20	105	–40
42	891 May 12	J	k'o	–	4	16	50	–20	0	+15
43	900 Feb./Mar.	C	k'o	–	5	17	00	+10	30	+30
44	911 May/Jun.	C	k'o	–	4	17	10	+15	35	+30
45	1006 Apr. 3	A,C,E,J	k'o	2 + years	1	15	10	–40	330	+15
46	1011 Feb. 8	C	k'o	–	4	19	20	–30	10	–20
47	1035 Jan. 15	C	'star'	–	5	1	20	+5	140	–55
48	1054 Jul. 4	C,J	k'o	22 months	1	5	40	+20	190	– 5
49	1065 Sep. 11	C	k'o	–	4	9	20	–25	255	+20
50	1069 Jul. 12	C	k'o	11 days	3	18	10	–35	0	–10
51	1070 Dec. 25	C	k'o	–	4	2	40	+ 5	165	–50
52	1073 Oct. 9	K	k'o	–	4	0	10	+ 5	105	–55
53	1074 Aug. 19	K	k'o	–	4	0	10	+ 5	105	–55
54	1138 Jun./Jul.	C	k'o	–	4	1	50	+20	140	–40
55	1139 Mar. 23	C	k'o	–	4	14	10	–10	335	+50
56	1163 Aug. 10	K	k'o	–	4	17	30	–20	5	+ 5
57	1175 Aug. 10	C	po	5 days	5	15	40	+50	80	+50
58	1181 Aug. 6	C,J	k'o	185 days	1	1	30	+65	130	0
59	1203 Jul. 28	C	k'o	9 days	3	17	10	–40	345	0
60	1224 Jul. 11	C	k'o	–	4	17	10	–40	345	0

(1)	(2)	(3)	(4)	(5)	(6)	(7)		(8)	(9)	(10)
61	1240 Aug. 17	C	k'o	—	4	17	10	-40	345	0
62	1356 May 3	K	k'o	—	4	5	50	+30	180	0
63	1388 Mar. 29	C	'star'	—	5	0	10	+20	110	-40
64	1399 Jan. 5	K	k'o	—	4	18	50	-20	15	-10
65	1404 Nov. 14	C	'star'	—	5	19	50	+30	65	0
66	1430 Sep. 9	C	'star'	26 days	4	7	30	+ 5	215	+10
67	1431 Jan. 4	C	'star'	15 days	4	4	50	-10	210	-30
68	1437 Mar. 11	K	k'o	14 days	3	16	50	-40	345	0
69	1460 Feb./Mar.	V	'star'	—	5	11	30	-15	275	+45
70	1572 Nov. 8	C,E,K	k'o	16 months	1	0	20	+65	120	0
71	1584 Jul. 11	C	'star'	—	5	16	00	-25	350	+20
72	1592 Nov. 28	K	k'o	15 months	1	1	20	-10	150	-70
73	1592 Nov. 30	K	k'o	4 months	1	0	50	+60	125	0
74	1592 Dec. 4	K	k'o	3 months	1	0	00	+60	115	0
75	1604 Oct. 8	C,E,K	k'o	12 months	1	17	30	-20	5	+ 5

Our revised catalogue of pre-telescopic galactic novae and supernovae which forms Table 3.1 contains information for seventy-five
candidates observed from earliest times until the beginning of
the telescopic era (AD 1609).   The columns give in order:

1. A reference number.
2. The Julian or Gregorian calendar date (the Gregorian
   calendar is used from AD 1582 September 15, the date
   of its inception).   Frequently only the month is given.
3. An initial letter denoting the place of observation
   (Arab Lands, China, Europe, Japan, Korea or Vietnam).
4. The type of star (k'o, po, hui, etc).
5. The duration of visibility, where known.
6. A classification number from 1 to 5, depending upon the
   reliability.

7 & 8. The approximate R.A. and dec. for epoch 1950.0,
       estimated from the oriental star maps of Ho Peng Yoke
       (1962, 1966).

9 & 10. The approximate galactic co-ordinates (l and b).

We have not considered it worthwhile to include the names of the
asterisms in or near which each star appeared.   This information
is of minor interest to the astrophysicist, whereas the sinolo-
gist would require further details.

Referring to Table 3.1, it is curious that no durations between
about twenty-five and fifty days are reported.   This provides a
means of selecting the most promising supernova candidates.   As
shown by Clark and Stephenson (1976), any star of duration less
than about forty days can be virtually ruled out as a supernova.
This is based partly on the evidence that there are only two
possible references to the long period variable star Mira Ceti in
Far Eastern history (AD 1070 and 1592) before its discovery in
Europe by David Fabricius in AD 1596.   This star has a period of
roughly eleven months.   At maximum it reaches a magnitude of
between +2 and +4 and at minimum descends to about +9.   It is
thus only visible to the unaided eye for about half of the time
and yet at a bright maximum it is the brightest star in its neighbourhood.   At a declination of $-3°$ it is very well placed for
general observation.   Taking conservative estimates of magnitude
+3 as the minimum brightness for detection and +5.5 as the lower
limit of unaided eye visibility, judging from the Type I and
Type II light curves of Barbon et al (1974a,b) a supernova would
take at least forty days, and possibly much longer, to decline
over this range of brightness.   There is the possibility that a
supernova might appear shortly before conjunction with the Sun
and thus be visible for only a few days.   However in Table 3.1
only a single k'o of specified but short duration appeared near
the sun.   This occurred in AD 1069, and as discussed in Chapter
4 there is no known supernova remnant in its vicinity.

From the above argument, we should expect to find any supernovae
among the longer duration stars (more than about fifty days),
while stars of short duration (less than about twenty-five days)
are not worth seriously considering as supernovae.   This, of

course, is not to imply that all stars of longer duration are
supernovae; we might expect to find both slow novae and comets
among this material.   However, there can be little doubt that
the short duration objects are a mixture of novae and comets
only.

Referring to the classification in column 5 of Table 3.1, stars
in classes 1 and 2 are all objects which were seen for more than
about fifty days.   Class 1 objects are all k'o for which there
is no hint of any tail or motion in the record.   Stars in class
2 are either downgraded k'o (because of possible reference to a
tail or motion) or stars of other type.   Basically class 3
objects are k'o of short but definite duration (up to a maximum
of about twenty-five days).   Class 4 stars are k'o of indefinite
duration or unspecified stars of known duration.   Finally,
objects in class 5 are unspecified stars of indefinite duration
or po of known duration.   An object is downgraded by 1 class if
there is any suggesting of a tail or motion.

It seems likely that most of the stars in class 3 were novae.
However, the proportion of comets in classes 4 and 5 may be very
significant.

Fig. 3.1 shows the positions of all of the stars in the catalogue
in terms of galactic co-ordinates (l and b).   The blank zone in
the lower right of the figure represents the area currently not
visible from 35°N latitude (the approximate location of the
various Chinese, Japanese and Korean capitals).   Stars of medium
to long duration (more than about fifty days) are represented by
shaded circles, others by open circles.

Outside of the blank area, the distribution in both l and b is
fairly isotropic.   This, of course, is what we would expect for
comets.   From the statistics of Galactic novae collected by
Payne-Gaposchkin (1957), these stars in general show a marked
concentration towards the galactic equator.   The distribution of
novae in galactic longitude is also very anisotropic, with a
pronounced peak in the direction of the galactic centre (l = 0°).
However, it is rather problematical to assess the expected
distribution of novae bright enough to attract the attention of
the unaided eye.   An indirect method is once more to make use
of the almost complete lack of pre-1596 sightings of Mira Ceti.
The maximum brightness of this star is about magnitude +2.   A
star brighter than +1 could scarcely be missed under normal
circumstances.   It would therefore appear that to stand a
reasonable chance of being discovered by the early Far Eastern
astronomers a maximum brightness of somewhere between these two
values would be required.   It seems appropriate to take the mean,
i.e. +1.5.

Assuming an average magnitude of -7.5 for galactic novae (Payne-
Gaposchkin, 1957) and allowing for interstellar absorption of
approximately 1 mag. per kpc (a smaller value would be appropriate for stars well off the plane) gives an average distance
for novae bright enough to be noticed of about 500 pc.   At such

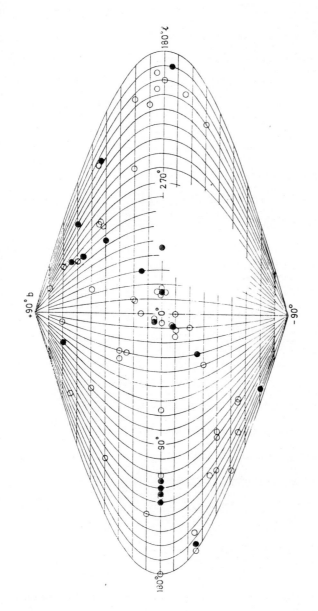

Fig. 3.1.   Galactic distribution of pretelescopic novae and supernovae.

a small mean distance we would expect the distribution of historical novae in galactic longitude to be almost uniform.

Again from Payne-Gaposchkin (1957), the average distance of novae from the galactic plane in the vicinity of the Sun is about 275pc. A fairly isotropic distribution in galactic latitude is thus also expected for historically recorded novae.

It is obviously not possible to give a useful estimate of the proportion of novae in the catalogue. All that can be done is to select the most probable supernovae and novae and leave the nature of the remaining objects in doubt. Because of their vast distance from us (the nearest historically observed supernova, which occurred in AD 1006, lies at a distance of 1 kpc), super- novae tend to lie close to the galactic equator. From the investigation by Clark and Stephenson (1976), the mean value of b for young (less than 2,000 years old) nearby (within about 10 kpc of the Sun) SNRs is only $2^{o}.3$. Independent evidence for super- novae belonging to the plane of a parent galaxy has been produced by Vettolani and Zamorani (1976) from an interpretation of the distribution of 132 supernovae detected in spiral and irregular galaxies. We thus expect to find galactic supernovae only among the new stars of long duration and low galactic latitude, (see also Chapter 4).

Of the temporary stars of short duration, only two are sure novae. The records of both the stars of AD 837 (A) and 1437 imply that the objects remained fixed for several days. The latter appeared close to the galactic equator, but the duration of fourteen days is much too short to make it worth considering as a possible supernova, particularly as the star was more than 90 degrees from the Sun and would thus be visible in a dark sky for several hours each night. Presumably it was a fast nova. The star of AD 1431 is a possible nova of the DQ Her. type (with a deep minimum) but the record is very brief.

Attention has sometimes been drawn to the possibility of two seemingly independent new stars representing separate outbursts of the same (recurrent) nova. In view of the poor positioning of such stars in general and, with a few notable exceptions, the marginal evidence that any particular star was a nova, no suggested recurrent novae among the pretelescopic objects can be taken seriously.

The twenty stars of duration in excess of about fifty days are summarized in Table 3.2, and their galactic distribution is shown in Fig. 3.2. The columns of Table 3.2 given in order are:

1. The reference number in Table 3.1
2. The year of occurrence.
3. The duration of visibility.
4. The galactic latitude.

Of the new stars in high galactic latitude ($|b| > 25^{o}$), the five objects of 5 BC, AD 61, 64, 247, and 402 may well have been

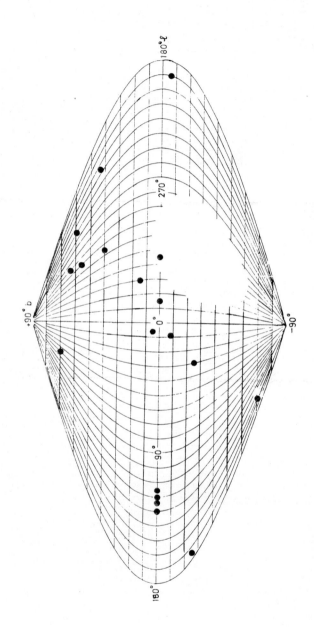

Fig. 3.2.   Galactic distribution of pretelescopic new stars of long duration.

REF. NO.	YEAR	DURATION	b
08	5 BC	70+ days	− 25°
09	AD 61	70 days	+ 70
10	64	75 days	+ 55
11	70	48 days	+ 45
17	185	20 months	0
19	247	156 days	+ 40
23	369	5 months	−
24	386	3 months	− 10
25	393	8 months	0
26	396	50+ days	− 25
27	402	2 months	+ 60
39	837 (B)	75 days	+ 65
45	1006	several years	+ 15
48	1054	22 months	− 5
58	1181	185 days	0
70	1572	16 months	0
72	1592 (A)	15 months	− 70
73	1592 (B)	3 months	0
74	1592 (C)	4 months	0
75	1604	12 months	+ 5

Table 3.2.   Temporary stars of medium and long duration.

comets since there is a possible allusion to motion or a tail.
The remaining high latitude stars are those of AD 70, AD 396,
AD 837(B), and AD 1592(A).   On the basis of high galactic
latitude alone, it is tempting to classify these events as novae;
there are certainly no SNRs at such high latitudes which could
possibly correspond (see Chapter 4).   In any case, even if they
were supernovae, on account of their high galactic latitude they
would have to be relatively nearby.   Extreme brightness would
thus be expected, but in no case is this reported.   There is
the possibility that the new star of AD 1592(A) was Mira Ceti.
Two maxima are seemingly recorded for the star, about a year
apart.   Mira has a period of roughly 11 months, but the recorded
position of the star is more than 10 degrees away.   A second
possible nova with a deep minimum in the light curve followed by
a secondary maximum (similar to the recent nova DQ Her. of 1934)
is the new star of AD 396.   This was originally seen for a
period of more than fifty days and was then recovered after a
lapse of three months.   However, as mentioned above, records
for this period tend to be very brief so that the true nature of
the object must be considered doubtful.   Indeed, in the absence
of additional information, the new stars of AD 70, 396, 837(B),
and 1592(A) can, at best, be assigned as probable novae.

In looking for possible supernovae, we are thus left with the new
stars of AD 185, 369, 386, 393, 1006, 1054, 1181, 1572, 1592(B),
1592(C), and 1604.   These will be considered further in the
latter half of the next chapter.

Chapter 4

THE SEARCH FOR THE REMNANTS
OF SUPERNOVAE

In Chapter 1, the four principal observable remnants of a supernova were given as a pulsar, an expanding optical nebulosity, a
region of radio emission, and an extended X-ray source. Of the
more than 100 objects within our own galaxy positively identified
as the remnants of supernovae, only the Crab Nebula and Vela
remnants are known definitely to display all four features. For
the remainder, the extended radio source is the most obvious
remnant, and appears to be common to all events; because of
obscuration effects, optical and X-ray emission may be detected
only for the remnants of supernovae which occurred relatively
nearby. The possible detection of a pulsar is subject to
observational limitations, and in addition it is by no means yet
certain that all supernova events produce an observable pulsar.

The radio remnants of supernovae came under investigation during
the very earliest days of radio astronomy after World War II, and
in fact the very first discrete stellar object (other than the
Sun) identified as a radio source was the Crab Nebula, the remnant
of the supernova of AD 1054. This identification was made by
Bolton (1948) using a 'sea interferometer' mounted on a cliff
over-looking the ocean near Sydney, Australia. That SNRs are
one class of galactic radio source was established almost from
the beginning of the science of radio astronomy, as was the nature
of the radio emission from such remnants.

The discrete celestial radio sources emit a smooth 'continuum' of
radio frequency energy (i.e. the emission intensity varies
smoothly, with emission frequency). In addition, some sources
show enhanced 'line' emission at certain characteristic frequencies. Continuum radio emission from our galaxy is known to
come from two principal types of extended object.

A. <u>Ionized Hydrogen Clouds</u>. The interstellar hydrogen in the
galaxy tends to be distributed in clouds, and this normally
remains in its neutral (H I) state. However if a hot stellar
source lies near or within such a cloud, ultra-violet radiation
from the source tends to ionize the cloud, separating the
hydrogen atoms into their constituent protons and electrons.
When an electron and proton in random thermal motion approach
each other at speed, they interact. The interaction is termed
"free-free", since the particles are unbound before and after
the interaction. The proton shows only a small deflection from
its original path, but the lighter electron completely changes
its direction emitting radiation known as <u>thermal radiation</u>, the
wavelength of which depends upon the particle velocities and the
distance at which they interact. In a hot gas the intensity of

(a)

(b)

Fig. 4.1. (a) Thermal radiation is produced by
the 'free-free' interaction of an electron and
a proton. The proton shows only a small deflection from its original path, but the lighter electron completely changes its direction emitting
radiation. (b) The radio-spectrum of the Orion
Nebula, a typical ionized hydrogen region.

the thermal radiation remains essentially constant with wavelength, although beyond a certain point the intensity decreases
with increasing wavelength (see Fig. 4.1).   This typical
'thermal spectrum' is unique to radio sources associated with
ionized hydrogen (H II) clouds within our galaxy.   An additional
feature of some such clouds is the line emission which may occur
from ionized hydrogen regions during a transient recombination of
an ion and electron (for example, the H109α recombination line
at a frequency of approximately 5,009 MHz).

B.   Supernova Remnants.   The earliest observations of the Crab
Nebula radio source showed that the intensity of the emitted
energy increased with increasing wavelength.   This is referred
to as non-thermal radiation to distinguish it from the thermal
radiation already described.   Radio sources later identified as
the remnants of Tycho's (AD 1572) and Kepler's (AD 1604) supernovae were also found to emit non-thermal radiation.   The
mechanism for producing such a spectral distribution remained,
for many years, one of the major mysteries of radio astronomy.
An explanation was eventually found by Shklovsky (1953), who
suggested that the emission was produced by the so-called
synchrotron mechanism.   The synchrotron is a type of particle
accelerator.   When in a synchrotron electrons are accelerated
to very high speed in a magnetic field, they radiate electromagnetic energy at a variety of wavelengths.   The mechanism is
the same in supernovae, where high energy electrons spiralling
along magnetic field lines radiate light, while those of low
energy radiate at radio wavelengths (see Fig. 4.2).   The
variation of emission intensity I with frequency v follows a
simple power law of the form $I(v) \sim v^{\alpha}$, with negative spectral
index α.   Apart from the unique spectral distribution, which is
related to the energy distribution of the electrons which produce
the emission, an additional characteristic of synchrotron
radiation is that it displays a high degree of polarization, a
property verified for the emission from SNRs (see Fig. 4.3).

Most extended non-thermal radio sources close to the galactic
plane have morphological and spectral characteristics similar to
the unambiguously identified galactic SNRs, and are consequently
themselves possible candidates.   However the recognition of a
non-thermal spectrum is not sufficient grounds for classifying a
radio source as a galactic SNR, since extragalactic radio sources
also display this characteristic.   Identification with optical
filamentary structure and/or a historically recorded supernova
may be possible for relatively nearby remnants.   In the absence
of such an identification, the final evidence sought is the
spatial distribution of the radio brightness of the source.
From a study of many radio remnants of supernovae observed with
both good sensitivity and high resolution, it is found that their
brightness can be modelled quite well by a uniform disc of
emission with a superimposed annulus of 'limb-brightening'.
Such a 'ring' source is indicative of a shell emitting region
centred on the site of the original supernova outburst.   Fig.
4.4 shows two examples of SNRs displaying this characteristic
peripheral emission.   It must be emphasised that these examples

(a)

(b)

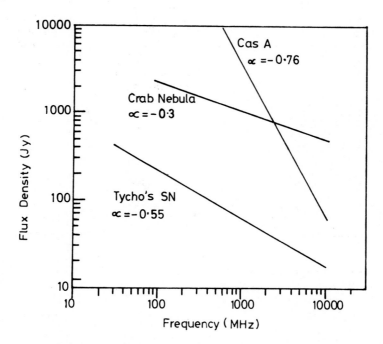

Fig. 4.2.   (a) A fast electron in a magnetic
field is constrained to move in a helical path
wrapped around the field lines, and emits non-
thermal radiation in the direction of motion.
(b) Representative continuum radio spectra of
three supernova remnants.

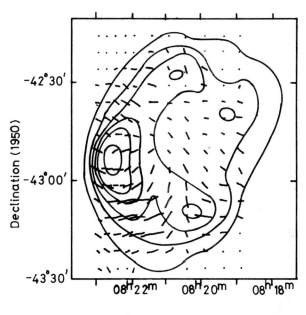

Right ascension (1950)

Fig. 4.3. A 2700 MHz map of the radiation from
the supernova remnant Puppis A, showing vectors
of polarization intensity. (From Milne 1971).

have been carefully chosen; many SNRs show only a part-ring or
broken-shell emission, while a few (notably the Crab Nebula radio
source) are amorphous displaying no limb-brightening, (more
later).

To briefly summarize, in attempting to classify a celestial radio
source as the remnant of a supernova one seeks (a) where possible,
an identification with an optical filamentary structure and/or an
historically recorded supernova, (b) a non-thermal spectrum and
some degree of polarization indicative of the synchrotron process,
and as a consequence of this property the absence of line emission
which characteristically accompanies free-free continuum radio
emission, (c) where resolution permits, the recognition of a
'ring' or 'part-ring' radio-brightness distribution, and (d)
proximity to the galactic plane.   On the basis of these classification criteria, the first comprehensive catalogues of galactic
SNRs were produced independently by Milne (1970) and Downes
(1971).   A few of the sources in these catalogues have now been
reclassified as ionized hydrogen clouds from improved observational data, and many new remnants have been identified by Clark,
Caswell, and Green (1973, 1975) in a survey of the southern
portion of the galactic plane made with the Mills Cross and Parkes

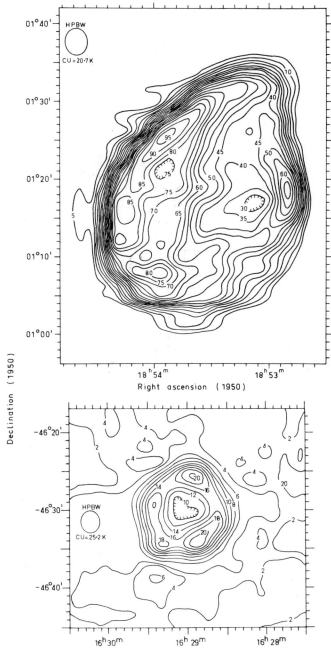

Fig. 4.4. Two examples of radio remnants displaying characteristic peripheral radio emission. (a) The SNR W44. (b) The SNR Kes 40. Both maps, at 408 MHz, were made with the Mills Cross radiotelescope of the University of Sydney.

radio telescopes in Australia .   The most recent catalogues (see
for example Green 1974, Clark and Caswell, 1976) now list about
120 galactic radio sources which are believed to be SNRs, and a
galactic SNR catalogue is reproduced here as Table 4.1.

The table is composed as follows: column 1 gives the galactic
source number of the SNR (galactic longitude followed by galactic
latitude), and column 2 other catalogue numbers or common names;
columns 3 and 4 give the right ascension and declination (epoch
1950.0) of the centroid of radio emission; column 5 gives the
angular diameter of the source measured from the highest-
resolution radio map available and referenced in column 6.
Column 7 gives the spectral index α.   The radio brightness
integrated over the source yields the total flux density, and
this is included in Column 8. (1 jansky (Jy) = 10^{-26} Wm^{-2} Hz^{-1}).

Woltjer (1970, 1972) has suggested that the dynamical evolution
of an SNR shell may conveniently be divided into four phases, and
this division has been adopted in most subsequent investigations.

Phase 1 (the free-expansion phase): The expansion of the supernova ejecta is essentially free following the supernova outburst
(possibly with continuing injection of energy from a pulsar), and
the appearance of the remnant depends on the initial conditions.
A shock front forms at the leading edge of the swept-up interstellar gas but the mass of this gas is insufficient to deceler-
ate the ejected matter noticeably.

Phase 2 (the adiabatic-expansion phase): The remnant is now
dominated by swept-up matter, but radiative losses are still
negligible and hence the total energy (kinetic plus thermal) is
conserved.   Shklovsky (1962) suggested that the expansion now
resembles that of an adiabatic blast wave created by releasing
energy at a point in a homogeneous gas.   This phase continues
only up to a time when radiation losses become important; thereafter the assumption of energy conservation is no longer valid.

Phase 3 (the isothermal-expansion phase): In this phase the
structure of the object changes, as radiative cooling losses
become significant.   The matter that passes through the shock
front cools rapidly and the density becomes high, with the formation of thin sheets or filaments which radiate in the optical and
ultra-violet ranges.   The remnant is now a rather thin shell
ploughing through the interstellar medium.   The thermal energy
is small, and the shell may be considered to be moving with
constant linear momentum.

Phase 4 (the extinction phase): When the expansion velocity of
the shell becomes comparable to the thermal or random motion in
the surrounding interstellar gas, the SNR gradually loses its
identity and merges with the interstellar medium.   It is of
interest only in so far as it contributes to the intensity of the
galactic background emission, and transfers kinetic energy to the
interstellar medium.

Table 4.1

A Catalogue of 120 Galactic SNRs

(1) Galactic Source Number	(2) Catalogue number or name	(3) Right ascension (1950.0) h m s	(4) Dec. ° '	(5) Angular diameter (arc. min)	(6) Ref.	(7) α	(8) 1000 MHz Flux density (Jy)
G180.0−1.7	S147	05 36 00	27 30	171	16	−0.5	64
G184.6−5.8	Crab Nebula	05 31 31	21 59	5.2	33	−0.3	996
G189.1+2.9	IC 443	06 14 00	22 35	40	19	−0.4	161
G193.3−1.5	PKS 0607+17	06 05 50	16 40	80	7		27
G205.5+0.2	Monoceros	06 35 00	06 30	253	7	−0.5	115
G206.9+2.3	PKS 0646+06	06 46 00	06 30	80	24		5
G260.4−3.4	Puppis A	08 21 00	−42 50	47	14	−0.48	129
G261.9+5.5	PKS 0902−38	09 02 22	−38 29	40	18		8
G263.9−3.3	Vela X, Y, Z	08 32 30	−45 35	256	7	−0.5	1469
G287.8−0.5		10 45 00	−59 23	<42	20		
G290.1−0.8	MSH 11−61A	11 00 52	−60 37	12.6	27	−0.55	68
G291.0−0.1	MSH 11−62	11 09 49	−60 22	10.0	27	−0.35	16
G292.0+1.8	MSH 11−54	11 22 22	−58 59	5.4	27	−0.41	15
G293.8+0.6		11 33 05	−60 36	9.0	5	−0.58	5
G296.1−0.7		11 48 15	−62 27	16.0	5	−0.7	4
G296.5+10.0	PKS 1209−52	12 07 00	−52 07	81	31	−0.45	57
G296.8−0.3		11 55 48	−62 18	14.9	4	−0.62	9
G298.5−0.3		12 09 58	−62 36	3.7	27	−0.36	5
G298.6+0.0		12 11 18	−62 18	8.3	27	−0.30	4
G299.0+0.2		12 15 05	−62 08	10.5	5	−0.39	9

(1)	(2)	(3)	(4)	(5)	(6)	(7)	(8)
G302.3+0.7		12 42 54	-61 51	16.5	5	-0.36	5
G304.6+0.1	Kes 17	13 02 35	-62 26	6.9	27	-0.48	14
G308.7+0.0		13 38 05	-62 01	7.3	5	-0.35	12
G309.2-0.6		13 43 00	-62 36	12.6	·5	-0.37	7
G309.8+0.0		13 47 03	-61 50	19.2	5	-0.51	17
G311.5-0.0		14 01 58	-61 43	3.9	27	-0.48	4
G315.4-0.3		14 32 00	-60 22	16.0	5	-0.47	10
G315.4-2.3	RCW 86	14 39 00	-62 17	39	4	-0.62	49
G316.3-0.0	MSH 14-57	14 37 43	-59 47	17.1	27	-0.32	28
G320.4-1.2	MSH 15-52	15 10 30	-59 00	25.8	27	-0.34	69
G321.9-0.3		15 16 45	-57 23	24.2	5	-0.34	13
G322.3-1.2		15 23 05	-57 56	5.8	4	-0.90	6
G323.5+0.1		15 25 05	-56 12	10.8	5	-0.41	3
G326.3-1.8	Kes 24	15 48 50	-56 00	36	6	-0.24	145
G327.1-1.1	MSH 15-56	15 50 35	-54 58	14.2	5	-0.36	8
G327.4+0.4	Kes 27	15 44 54	-53 39	21.0	4	-0.61	34
G327.6+14.5	PKS 1459-41	14 59 30	-41 45	34	23	-0.57	19
G328.0+0.3		15 49 33	-53 19	6.4	27	-0.55	3
G328.4+0.2	MSH 15-57	15 51 45	-53 08	4.0	27	-0.24	16
G330.0+15.0	Lupus Loop	15 09 00	-39 00	368	23		284
G330.2+1.0		15 57 20	-51 25	8.3	5	-0.30	7
G332.0+0.2		16 09 23	-50 49	12.0	27	-0.44	10
G332.4+0.1	MSH 16-51	16 11 38	-50 32	13.2	27	-0.51	25
G332.4-0.4	RCW 103	16 13 48	-50 54	9.4	27	-0.55	28
G335.2+0.1		16 23 50	-48 36	18.6	5	-0.46	18

Table 4.1. Continued

(1)	(2)	(3)	(4)	(5)	(6)	(7)	(8)
G336.7+0.5		16 28 30	-47 13	9.9	27	-0.37	7
G337.0-0.1	CTB 33	16 32 08	-47 30	7.6	27	-0.47	17
G337.2-0.7		16 35 45	-47 44	3.9	5	-0.67	2
G337.3+1.0	Kes 40	16 29 05	-46 29	11.8	4	-0.49	16
G337.8-0.1	Kes 41	16 35 15	-46 53	10.6	27	-0.51	16
G338.2+0.4		16 34 40	-46 16	11.7	27	-0.42	2
G338.3-0.1		16 37 25	-46 27	8.2	27	-0.66	7
G338.5+0.1		16 37 20	-46 12	12.4	27	-0.33	27
G339.2-0.4		16 41 50	-46 03	9.9	5	-0.20	6
G340.4+0.4		16 43 00	-44 34	6.4	5	-0.41	6
G340.6+0.3	MSH 16-48	16 44 05	-44 30	4.9	5	-0.36	5
G341.9-0.3		16 51 20	-43 54	6.1	3	-0.59	4
G344.7-0.1		17 00 15	-41 37	7.8	5	-0.51	3
G346.6-0.2		17 06 45	-40 06	8.0	5	-0.49	10
G348.5+0.1	CTB 37A	17 11 12	-38 26	8.0	6	-0.33	72
G348.7+0.3	CTB 37B	17 10 45	-38 06	5.1	6	-0.30	26
G349.7+0.2		17 14 37	-37 23	1.7	28	-0.49	20
G350.0-1.8		17 23 45	-38 20	28.9	5	-0.51	31
G350.1-0.3		17 17 40	-37 24	4.1	5	-0.73	6
G351.2+0.1		17 19 00	-36 09	6.2	5	-0.38	6
G352.7-0.1		17 24 20	-35 05	6.4	5	-0.57	6
G355.9-2.5		17 42 30	-33 43	11.2	5	-0.51	8
G357.7-0.1	MSH 17-39	17 37 04	-30 56	5.2	28	-0.43	37
G4.5 +6.8	Kepler's SN	17 27 41	-21 26	3.2	15	-0.58	20
G5.3 -1.0	A4	17 58 30	-24 50	15.0	6	-0.2	32

(1)	(2)	(3)	(4)	(5)	(6)	(7)	(8)
G 6.4-0.1	W28	17 57 35	-23 27	49	27	-0.38	327
G 7.7-3.7		18 14 15	-24 05	19.5	Unpublished	-0.25	10
G10.0-0.3		18 05 42	-20 25	6.0	27	-0.34	2
G11.2-0.3		18 08 31	-19 26	4.2	28	-0.56	22
G11.4-0.1		18 07 50	-19 06	7.0	5	-0.48	6
G12.0-0.1		18 09 16	-18 38	5.4	5	-0.71	3
G15.9+0.2		18 16 07	-15 02	5.0	5	-0.56	5
G18.8+0.3	Kes 67	18 21 00	-12 25	15.0	6	-0.36	27
G21.8-0.6	Kes 69	18 30 05	-10 09	22.8	27	-0.54	68
G22.7-0.2		18 30 35	-09 13	24.6	27	-0.57	32
G23.3-0.3	W41	18 31 40	-08 51	21.0	27	-0.50	59
G23.6+0.3		18 30 25	-08 14	7.2	27		5
G24.7+0.6		18 31 30	-07 08	14.0	27	-0.59	16
G24.7-0.6		18 36 00	-07 37	14.6	5	-0.49	8
G27.4+0.0	Kes 73	18 38 36	-04 59	4.4	3	-0.45	3
G29.7-0.2	Kes 75	18 43 48	-03 02	2.4	27	-0.71	10
G31.9+0.0	3C 391	18 46 47	-00 59	4.8	28	-0.57	51
G32.0-4.9	3C 396.1	19 04 30	-03 06		2	-0.5	
G32.8-0.1	Kes 78	18 48 55	-00 13	17.2	4	-0.20	11
G33.6+0.1	4C 00.70	18 50 05	00 37	9.2	4	-0.60	21
G34.6-0.5	W44	18 53 35	01 17	27.2	6	-0.28	232
G39.2-0.3	3C 396	19 01 35	05 22	6.6	28	-0.49	19
G39.7-2.0	W50	19 09 15	05 12	>30	6		>25
G41.1-0.3	3C 397	19 05 08	07 04	3.6	4	-0.49	19
G41.9-4.1	PKS 1920+06	19 20 00	06 00	164	24	-0.5	19

(1)	(2)	(3)	(4)	(5)	(6)	(7)	(8)
G43.3 −0.2	W49B	19 08 43	09 01	4.2	28	−0.47	35
G46.8 −0.3		19 15 45	12 05	15.6	4	−0.42	14
G47.6 +6.1	CTB 63	18 54 00	15 45	83.4	24	−0.5	
G49.2 −0.5	W51	19 21 30	13 57	26.6	26	−0.25	160
G53.7 −2.2	3C 400.2	19 36 15	17 07	26.6	6	−0.32	9
G54.4 −0.3	HC 40	19 31 00	18 55	43	30	−0.49	27
G55.7 +3.4		19 19 40	21 37	17.4	12	−0.5	1
G65.7 +1.2	DA 495	19 50 05	29 17	18.6	32	−0.64	6
G74.0 −8.6	Cygnus Loop	20 49 00	30 30	180	21	−0.45	211
G74.9 +1.2	CTB 87	20 14 05	37 04	6.5	11	−0.47	9
G78.1 +1.8	DR 4	20 20 45	40 04	12.2	17	−0.65	55
G82.2 +5.4	W63	20 17 15	45 25	77	30	−0.7	133
G89.0 +4.7	HB 21	20 43 20	50 29	100	22	−0.40	259
G93.2 +6.7	4C(T) 55.38.1	20 51 00	55 07	25.0	Unpublished	−0.37	8
G93.6 −0.2	CTB 104A	21 26 50	50 33	61	30	−0.35	37
G94.0 +1.0	3C 434.1	21 23 30	51 40	26.6	30	−0.53	12
G111.7−2.1	Cas A	23 21 11	58 33	4.2	25	−0.76	3390
G117.3+0.1	CTB 1	23 56 40	62 09	30	32	−0.58	8
G119.5+10.0	CTA 1	00 02 35	72 20	130	1	−0.2	41
G120.1+1.4	Tycho's SN	00 22 37	63 52	7.9	29	−0.55	56
G130.7+3.1	3C 58	02 01 53	64 35	5.2	13	−0.1	35
G132.4+2.2	HB 3	02 15 00	62 25	72	30	−0.52	62
G160.5+2.8	HB 9	04 57 00	46 36	139	8	−0.44	135
G166.0+4.3	VRO 42.05.01	05 22 50	42 52	44	9	−0.40	7
G166.2+2.5	OA 184	05 15 30	41 47	76	10	−0.55	10

References for Table 4.1.

1. Caswell (1967)
2. Caswell (1970a)
3. Caswell and Clark (1975)
4. Caswell, Clark and Crawford (1975)
5. Clark, Caswell and Green (1973, 1975)
6. Clark, Green and Caswell (1975)
7. Day, Caswell and Cooke (1972)
8. Dickel and McKinley (1969)
9. Dickel, McGuire and Yang (1965)
10. Dickel and Yang (1965)
11. Duin et al. (1975)
12. Goss and Schwarz (1971)
13. Goss, Schwarz and Wesselius (1973)
14. Green (1971)
15. Gull (1975)
16. Haslam and Salter (1971)
17. Higgs and Halperin (1968)
18. Hill (1967)
19. Hill (1972)
20. Jones (1973)
21. Keen et al. (1973)
22. Kundu (1971)
23. Milne (1971a)
24. Milne and Hill (1969)
25. Rosenberg (1970)
26. Shaver (1969)
27. Shaver and Goss (1970a)
28. Slee and Dulk (1974)
29. Strom and Duin (1973)
30. Velusamy and Kundu (1974)
31. Whiteoak and Gardner (1968)
32. Willis (1973)
33. Wilson (1970)

Analytical solutions are available for the various phases.
Rosenberg and Scheuer (1973), Chevalier (1974), and others, have
attempted numerical computation of the full dynamical development;
this work has emphasized the uncertainties of assigning observed
SNRs to one of the above phases, since the transitions between
them are complex and with unknown time scales.   Nevertheless,
division into the four phases provides a useful first approximation for a study of SNR evolution.

Although the spectrum and polarization of the radio emission from
SNRs confirm that it is synchrotron radiation, there is no single
completely satisfactory theory yet for the origin and evolution
of the particles and fields.   There is increasing evidence that
these might be dominated by quite distinct physical processes at
various stages of a remnant's evolution.

Gull (1973a) has suggested that during Phase 1, instabilities at
the interface between the ejecta and the interstellar medium
result in a region of enhanced magnetic field which gives a shell
source of non-thermal radiation as relativistic particles diffuse
into the interstellar medium;   during this phase the radio
emission is expected to rise to a maximum and subsequently decay.
This model is found to describe well the radio properties of such
young SNRs as Cas A and the remnant of Tycho's supernova (Gull
1973b).

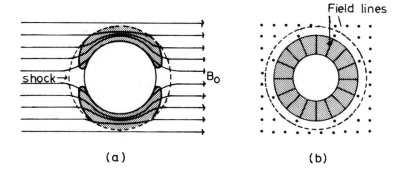

(a)   (b)

Fig. 4.5.   The evolutionary model of van der Laan.
(a) The interstellar magnetic field B_O perpendicular to the line of sight. The region of
maximum radio emission, shown shaded, appears as
two 'lobes' containing the compressed magnetic
field.
(b) The interstellar magnetic field along the
line of sight. The region of maximum emission is
a ring containing the transverse component of the
distorted magnetic field.
(From Whiteoak and Gardner, 1968).

For older remnants the interstellar magnetic field and relativistic cosmic ray electrons are expected to give rise to non-
thermal radiation after compression at the shock front, as in the
model of van der Laan (1962).   The model is illustrated in Fig.
4.5. Fig. 4.5(a) shows the case where the interstellar magnetic
field is perpendicular to the line of sight.   The maximum radio
emission then results from the two 'lobes' containing the
compressed magnetic field.   Many SNRs show this characteristic
double-lobe structure, and an example is shown as Fig. 4.6.
Measurements of polarization direction (always perpendicular to
the field orientation) would then be expected to be radial.   Fig.
4.5(b) shows the case where the interstellar magnetic field is
exactly along the line of sight.   Here synchrotron emission is
observed only where the distorted field has a transverse (i.e.
apparently radial) component, so that the radio emission would be
a complete 'ring' with zero intensity at the centre and the
direction of polarization circumferential.

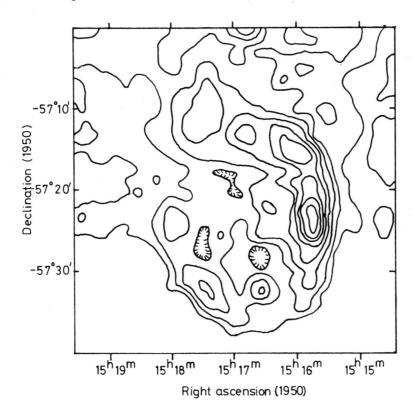

Fig. 4.6. A 408 MHz map of the remnant G321.9-0.3,
displaying the characteristic double-lobe structure of an SNR viewed across the interstellar
magnetic field.

The distribution of SNRs shows a concentration close to the galactic plane, and the majority lie within little more than the solar distance from the galactic centre.   A source count from the catalogue of SNRs gives the following distribution in galactic latitude b: 75 sources for $|b| < 1^{\circ}$; 34 sources for $1^{\circ} \leqslant |b| < 5^{\circ}$; 7 sources for $5^{\circ} \leqslant |b| < 10^{\circ}$;   and only 4 sources for $|b| \geqslant 10^{\circ}$. The highest known value of b for a young galactic SNR is 15°. Although most galactic SNR surveys have been restricted to within a few degrees of the plane, the concentration of remnants towards the plane is not believed to be a consequence of this;   indeed, Henning and Wendker (1975) have concluded from a search of available high-latitude radio data that there are no additional high-latitude sources which would definitely be assigned as SNRs. The existence of a few SNRs at large distance, say >300 pc, from the plane (for example, the remnant of the supernova of AD 1006 - more later) negates the suggestion that only supernovae occurring near the galactic plane leave observable remnants.

The first celestial X-ray source was discovered in 1962.   The present X-ray catalogues contain over 200 objects, several of which are SNRs.   X-rays from SNRs result from thermal radiation and line emission from the heated interstellar material swept up by the expanding blast wave;   in addition, synchrotron radiation is recognised as the principal source of X-ray emission in the Crab Nebula, although this X-ray emission mechanism has not been confirmed for any other SNR.

Following the free-expansion phase, the leading edge of the expanding shell of ejecta experiences significant deceleration with resulting increased density.   The faster material following enters this denser region from behind, with resultant heating of the gas and forming a second shock front behind the outer dense shell.   In the reference frame of the expanding shell this second shock front appears to be moving inward;   hence it is usually referred to as the 'reverse shock wave'.   McKee (1974) has suggested that the heated gas behind the reverse shock wave may be the source of soft thermal X-ray emission in young SNRs, while harder thermal X-ray emission might be detected behind the leading edge shock wave where the gas is heated to temperatures in the range $10^{7} - 10^{9}$ K.

During the adiabatic phase of a remnant's expansion, the velocity of the shock wave falls as the mass of swept up interstellar matter increases.   The temperature of gas behind the leading edge shock wave falls to about $10^{6} - 10^{7}$ K, and the X-ray emission is shifted to longer wavelengths with increased intensity due to the larger mass of radiating hot gas.   Concentration of the hot gas in an expanding shell results in peripheral brightening of the source (see Fig. 4.7), while inhomogeneities in the interstellar medium are usually invoked to explain the 'patchiness' of the X-ray emission with bright regions resulting from the encounter of the expanding shock wave with local interstellar density enhancements.

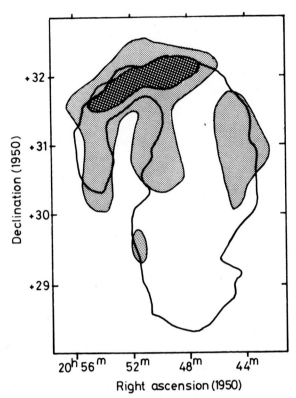

Declination (1950)

Right ascension (1950)

Fig. 4.7. The Cygnus Loop, showing the outline of the region of continuum radio emission
with the region of X-ray emission shaded ;
(a simplified version of a diagram from Seward
et al, 1976).

At the present time eight nearby SNRs have been definitely
confirmed to be X-ray emitters.   X-ray emission for the Crab
Nebula is recognised to be synchrotron in nature, with characteristic non-thermal spectrum and significant polarization.   At
X-ray wavelengths, the young SNRs of Tycho's supernova and Cas A
both exhibit thermal spectra which may be explained by the
'reverse shock wave' model.   The remnant of the supernova of
AD 1006 shows weak thermal emission, while the older remnants
Vela, Puppis A, IC443, and the Cygnus Loop all display more
intense thermal X-ray emission in accordance with predictions
from the adiabatic expansion model.   For recent reviews of the
X-ray remnants of supernovae, see Charles and Culhane (1975), and
Clark and Culhane (1976).

While the early chapters of the history of X-ray astronomy have
been dominated by the spectacular discoveries relating to the
existence of neutron stars and black holes in binary systems, it
is our belief that the later chapters will reveal great advances
in astrophysics achieved through the observation of the X-ray

SNRs.   The X-ray remnants of supernovae appear to be amongst the
most valuable of astronomical antiquities, capable of revealing
information on the initial supernova outburst, the evolution of
the remnants, and the structure and composition of the interstellar medium.

There are three main classes of optical emission nebulae in the
galaxy, although the atomic processes that produce the optical
emission for each of these three classes are basically the same.
Ionized hydrogen clouds represent the first class.   As mentioned
earlier, the ultraviolet emission from a hot central star will
ionize the cloud.   Transient recombination of a proton and an
electron can then take a variety of forms.   An electron can be
captured at any energy level of the atom, and then falls rapidly,
via a variety of possible intermediate states, to the ground
state.   The transition from the hydrogen atom's third energy
level to the second energy level produces the most intense visible
radiation, at a wavelength of 6,563 angstroms - the so-called $H\alpha$
emission.   The second class of emission nebulae are the planetary
nebulae.   The shell of gas emitted by a nova may be ionized by
emission from the post-nova.   Recombination in the expanding
shell is then evidenced by emission at wavelengths characteristic
of the chemical elements in the shell.   The third main class of
emission nebulae are the optical remnants of supernovae, where
the source of energy is not necessarily the photons from a central star.   In the case of the Crab Nebula, ionization is produced by synchrotron radiation.   However for the majority of
optical remnants, the ionization is believed to result from the
interaction of the expanding shock wave with inhomogeneities in
the interstellar gas;   the light we observe is then generated
behind the shock wave, where the ions and electrons are cooling
and recombining.   The relative abundances of different elements
in a remnant can then be interpreted in terms of the strengths of
its various emission lines.

Of the 120 galactic supernova remnants, only 24 display any
optical emission - the most recent catalogue of these optical
remnants is that of van den Bergh, Marscher, and Terzian (1973).
The optical remnants of supernovae, the most spectacular of which
include S147 (see Plate 6), the Crab Nebula, IC443, Vela, and
the Cygnus Loop, are amongst the most beautiful nebulosities in
the skies.

As noted in Chapter 1, only supernovae which occurred on the
near-side of the Galaxy could be expected to have been observed
from the Earth ;   the brightness of an outburst then gives a
crude estimate of its distance.   In looking for the probable
remnants of the historical supernovae, one therefore requires
distance estimates for the SNRs.   Such estimates may be obtained
in a variety of ways, but only the two most common techniques
will be described briefly here.

The most reliable technique for SNR distance determination uses
the absorption of the continuum radio emission from the SNR by
the neutral interstellar hydrogen (H I) along the line-of-sight.

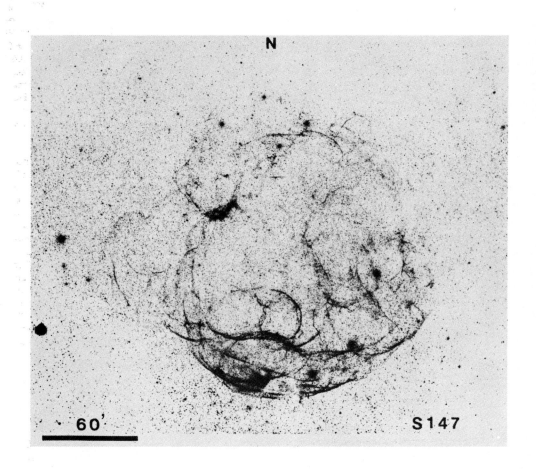

Plate 6.   The optical remnant S147.

The 21 cm (1420 MHz) spectral line from neutral hydrogen was first detected in emission by Ewen and Purcell (1951), and in absorption by Hagen and McClain (1954). The Doppler shift of the line observed in various directions gives the velocity with which the neutral hydrogen is moving towards or away from us, and with a suitable frequency (velocity) - distance model radio astronomers have been able to derive from such observations a picture of the spiral structure of our Galaxy (see for example Kerr, 1968). For neutral-hydrogen measurements in the direction of certain SNRs, well-defined absorption minima are exhibited; these are interpreted as being due to neutral-hydrogen clouds in the spiral arms along the line-of-sight. The positions of the intervening spiral arms for a particular SNR may be determined from a reference profile slightly displaced from the direction of the SNR, and the absorption feature corresponding to the most distant cloud then gives the minimum distance to the SNR. This H I absorption technique has been successfully applied to give minimum distances to about 30 galactic SNRs (see for exámple Caswell et al, 1976).

In the second technique, a distance for an SNR may be inferred from comparing its radio surface brightness (Σ) and angular diameter with those for SNRs with well determined values. (The mean surface brightness is defined as the integrated flux divided by the angular area of the source, and is measured in units of $Wm^{-2} Hz^{-1} sr^{-1}$). It has long been recognised (e.g. Shklovsky 1960 a, b) that the mean surface brightness of an SNR at a particular radio frequency is an extremely valuable observational parameter because it is distance-independent, and to a first approximation an intrinsic property. After reaching a maximum shortly after the birth of the supernova, Σ may be expected to decrease monotonically with time. (It has been conjectured that in later phases, as different factors become important, an increase might occur; however there is no real observational evidence for this). The outer diameter D of the expanding SNR will <u>increase</u> monotonically with time. Using SNRs of known distance we may then plot their distribution in the Σ - D plane, when it is found that they appear to follow a common evolutionary track (see for example Clark and Caswell, 1976). The expected linear diameter of an SNR of unknown distance but with measured surface brightness can then be found from the Σ - D plot, and a distance thus estimated from the measured angular diameter of the source. The Σ - D approach is at present an empirical one with little theoretical foundation. The limits of the technique are discussed by Clark and Caswell (1976); despite these limitations Σ - D distance estimates are believed accurate to within $\pm$ 30% for the majority of SNRs.

The catalogue of radio remnants of Clark and Caswell (1976) has been searched to see which sources, being of small diameter and high surface brightness, might be young enough to have originated in supernovae that have occurred within, say, the last two thousand years and which therefore might have been historically recorded. The variation of remnant diameter with time has not been well understood in the past because of the shortage of reliable age calibrators, although Clark and Caswell (1976), on

the basis of various assumptions detailed in the original reference, have derived the empirical relationship -

$$D \simeq 0.9 \ t^{2/5}$$

from a statistical analysis of their catalogue, where D is the
remnant diameter in parsec, and t is the time in years since
the explosion. (This diameter - time relationship will be
further investigated in Chapter 12). From the equation, an age
of 2000 years corresponds to a remnant of diameter about 20 pc
(with large uncertainty).

We have previously noted that a new star is most unlikely to be
discovered with the unaided eye if it is fainter than about
magnitude +3. As outlined in Chapter 1, a supernova near the
plane will reach magnitude +3 only if it occurs at a distance of
less than about 7 to 8 kpc. In our search for 'young' SNRs of
supernovae which might have been recorded historically, we therefore limited ourselves to those with estimated distances less
than 10 kpc, well beyond the expected limit.

The SNRs from the Clark and Caswell catalogue which meet the
above mentioned criteria (distance ⩽ 10 kpc, diameter ⩽ 20 pc)
are listed in Table 4.2. The Table is composed as follows :
Column 1 gives the Galactic source number of the SNR, and a common
name or alternative catalogue number; column 2 gives its declination (epoch 1950.0) - important in considering a possible historical detection; columns 3 and 4 give source distance and
diameter estimates found by the technique given in column 5 -
where such estimates are lower limits, the Σ - D parameters for
the source are also given in parentheses; the references for
these data are given in column 6. The sources are also shown
plotted (and identified by Table number, or year of outburst for
the well-established supernovae) on the simplified 'bird's-eye'
view of the Galaxy in Fig. 4.8. SNRs with lower limits for
distances are indicated by arrows directed away from the lower
limit value.

The first source in Table 4.2. is the well known remnant of
AD 1054 (to be discussed in detail in Chapter 8). Sources 2,
3, and 4 lie so far south as to reduce the likelihood of their
detection from Europe, the Middle or Far East. Source 5 will
be discussed in Chapter 5 as the probable remnant of the supernova
of AD 185. Source 6 is at an extreme southern declination, and
has a lower limit diameter which may imply an age > 2000 years,
Source 7 is the remnant of the supernova of AD 1006, and will be
discussed in Chapter 7.

The Historical Supernovae

Table 4.2

Young SNRs within 10 kpc of the Sun

	(1) Galactic source number (Other name)	(2) Declination (1950.0)	(3) Distance (kpc)	(4) Mean linear diameter (pc)	(5) Notes	(6) References.
1.	G184.6-5.8 (Crab Nebula)	+21°59'	2.2	3.0	A	7
2.	G290.1-0.8 (MSH 11-61A)	-60°39'	≥3.4 (5.8)	≥12.5 (21.1)	B	2
3.	G292.0+1.8 (MSH 11-54)	-58°59'	≥3.7 (13.4)	≥5.8 (21.0)	B	1
4.	G304.6+0.1 (Kes 17)	-62°26'	9.7	19.5	B	1
5.	G315.4-2.3 (RCW 86)	-60°22'	2	22	C	8
6.	G326.3-1.8 (MSH 15-56)	-56°02'	≥1.5 (3.2)	≥15.7 (33.4)	B	1
7.	G327.6+14.5 (SN 1006)	-41°45'	1.3	12.8	C	5
8.	G332.4-0.4 (RCW 103)	-50°58'	3.3	9.0	B	1

9.	G348.5+0.1 (CTB 37A)	-38°26'	≥6.7 (7.1)	≥15.6 (16.4)	B	
	G348.7+0.3 (CTB 37B)	-38°06'	≥6.7 (11.6)	≥9.7 (17.2)	B	1
10.	G357.7-0.1 (MSH 17-39)	-30°56'	≥6 (9.8)	≥9.1 (14.9)	B	6
11.	G4.5+6.8 (Kepler)	-21°26'	10	9.3	C	4
12.	G11.2-0.3	-19°26'	≥5 (12.1)	≥6.1 (14.8)	B	6
13.	G29.7-0.2 (Kes 75)	-03°02'	≥6.6 (17.9)	≥4.6 (12.5)	B	1
14.	G33.6+0.1	+00°37'	≥7 (9.4)	≥18.7 (25.1)	B	1
15.	G41.1-0.3 (3C397)	+07°04'	≥7.5 (13.7)	≥7.9 (14.3)	B	6
16.	G111.7-2.1 (Cas A)	+58°33'	3.3	3.7	B,C	4
17.	G120.1+1.4 (Tycho)	+63°52'	≥6 (7.2)	≥13.8 (16.5)	B	3,9
18.	G130.7+3.1 (3C58)	+64°35'	≥8 (11.1)	≥12.1 (16.8)	B	3,9

Notes : A : optical radial velocity and proper motion
B : H I absorption, kinematic distance
C : From optical and other considerations

References to Table 4.2

1. Caswell et al (1976
2. Goss et al (1972)
3. Goss et al (1973)
4. Ilovaisky and Lequeux (1972)
5. Minkowski (1966)

6. Radhakrishnan et al (1972)
7. Trimble (1968)
8. Westerlund (1969)
9. Williams (1973)

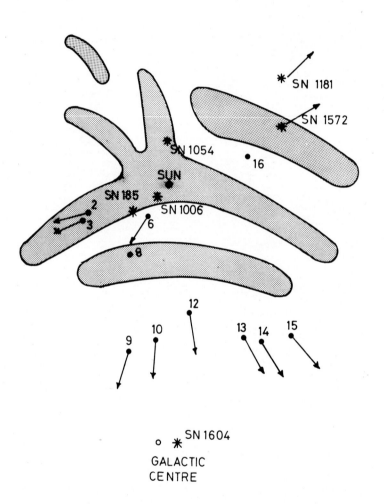

Fig. 4.8. A "bird's-eye view" of the Galaxy, showing the estimated positions of the SNRs from Table 4.2.

Source 8 has a well established diameter which would suggest it being a young SNR, and a distance at which a historical observation might have been expected although again the southerly declination might have reduced the likelihood of it being sighted from the northern hemisphere - this possibility is discussed in more detail later. Source 9 (in fact probably <u>two</u> adjacent SNRs) will be discussed in Chapter 6 as the possible remnant of the proposed supernova of AD 393. Source 10 is close to the galactic centre and lying in the plane, so that a visual detection would have been unlikely. Source 11 is the remnant of the supernova of AD 1604 (see Chapter 11). Sources 12, 13 and 15 all have only minimum H I absorption kinematic distance estimates, but large (>10 kpc) Σ - D distances making them poor candidates for detection. However in Chapter 6 we introduce the possibility that source 12 is the remnant of a supernova in AD 386. The Σ - D diameter for source 14 exceeds the 20 pc (equivalent to age 2000 years) limit discussed above. Source 16, Cas A, is almost certainly the remnant of a young, nearby but undetected, supernova. Source 17 is the remnant of the supernova of AD 1572 (see Chapter 10) and source 18 is probably the remnant of the supernova of AD 1181 (see Chapter 9). Thus in Table 4.2, apart from the proposed remnants of the long-duration new stars (all believed to be supernovae), only source 8 presents itself as a likely candidate for historical detection, with sources 12 through 15 less likely candidates. There is no source in the Table which could correspond to the event of AD 1069 to which we drew attention in Chapter 3.

Of the long duration new stars of interest listed in Chapter 3, only those of AD 369, and AD 1592 B and C remain to be accounted for. The position of the star of AD 369 is so poorly recorded that we cannot even deduce an approximate galactic latitude. The probable declination of this star was around +65°, but the right ascension is in considerable doubt. However north of +50° there are no 'young' SNRs, other than G111.7-2.1 (Cas A), G120.1+1.4 (AD 1572), and G130.7+3.1 (AD 1181). We thus conclude that the star was probably a slow nova. Both stars of AD 1592 appeared in Cassiopeia, the former in the neighbourhood of α Cas and the latter close to β Cas. This area has been well surveyed for SNRs, and none are known apart from Cas A. Brosche (1967) and Chu Sun-il (1968) independently suggested that Cas A (G111.7-2.1) is the remnant of the star AD 1592C, but the preferred age of the remnant is much less than this (see for example Gull 1973b). It seems likely that both stars were slow novae.

For most of the time since AD 1280 the Chinese capital has been Peking ($\approx$40° North). From this latitude a supernova south of declination 50° would not have been sighted. This is a possible explanation for the non-detection of the star which produced the young SNR G332.4-0.4, previously noted as a likely candidate for historical observation because of its proximity to the Earth. Alternatively the supernova may have occurred near the time of conjunction with the Sun, reducing the likelihood of its detection.

In the following chapters we present detailed discussions of the
8 historically-recorded new stars of long duration (i.e. of
AD 185, 386, 393, 1006, 1054, 1181, 1572 and 1604) proposed, with
varying degrees of certainty, as possible supernovae.

Chapter 5

THE GUEST STAR WITHIN THE
SOUTHERN GATE

The new star of AD 185 is the earliest recorded for which there
are any grounds for supposing it to be a supernova. The star
was reported only in China, and all we know about it is contained
in a single text.

No reference to the guest star is expected from Japan or Korea
at such an early period. In the former country we are still in
the legendary era. A number of astronomical records from Korean
history around this time are accessible (in the Samguk Sagi), but
these are of doubtful reliability; in any event, there is no
mention of the star. As it was only visible south of about
latitude 35°N (it appeared in the constellation of Centaurus),
Europe can be ruled out as a possible source of observations.
There seems a possibility that the star might have been carefully
observed in Alexandria (31°) , the home of the great astronomer
Claudius Ptolemy half a century before, but, if so, no account of
it has survived the ravages of time.

The solitary Chinese description of the guest star is to be
found in part of the astronomical treatise (Chapter 22) of the
Hou-han-shu. A translation of the text is as follows :-
"2nd year of the Chung-p'ing reign period (of Emperor Hsiao-ling),
10th month, day kuei-hai, a guest star appeared within (chung)
Nan-mên. It was as large as half a mat; it was multicoloured
(lit. 'It showed the five colours') and it scintillated. It
gradually became smaller and disappeared in the 6th month of
the year after next (hou-nien). According to the standard
prognostication this means insurrection. When we come to the
6th year, the governor of the metropolitan region Yüan-shou
punished and eliminated the middle officials. Wu-kuang attacked
and killed Ho-miao, the general of chariots and cavalry, and
several thousand people were killed". The date of appearance
of the guest star corresponds to AD 185 December 7.

Before commenting on the above record, it is appropriate to
discuss its historical and astronomical background.

The various treatises of the Hou-han-shu were compiled by
Szŭ-ma Piao, a scholar and member of the imperial family who
lived between AD 240 and 305 (see for example Han,1955).
Szŭ-ma Piao was attached to the imperial library and would thus
presumably have free access to the official documents of the
previous dynasty. The remainder of the history (annals and
biographies) was written more than a century later by Fan-yeh
(AD 398-445) and the whole was put together about AD 510. It is
unfortunate that only very minor sections of the Hou-han-shu

(including the astronomical records) have ever been translated into English.

Emperor Hsiao-ling, who was on the throne when the guest star occurred, reigned from AD 168 to 189. His capital was Lo-yang, which had been the seat of government since the beginning of the Later Han dynasty. As discussed in Chapter 2, we can confidently assume that the observations of the star which are incorporated in the extant record were made at Lo-yang ($34^\circ.7$ N) by the imperial astronomers. Hsiao-ling was the penultimate ruler of the dynasty, which at the time of his reign was already on the verge of collapse. The year before the star appeared, a major rebellion occurred in the northern provinces. This was one of the many calamities which brought about the downfall of the Han dynasty. The star thus appeared at a very critical time in Chinese history, and, like other celestial portents, must have been regarded as a harbinger of misfortune.

The imperial astronomers at this epoch were skilled observers as their records contained in Chapters 21, 22 and 28 of the Hou-han-shu testify. As an example, we might cite the following description of a meteor which occurred in the year AD 178: "1st year of the Kuang-ho reign period (of Hsiao-ling), 4th month, (day) kuei-ch'ou, a meteor (liu-hsing) trespassed against the 2nd star of Hsüan-yüan. It travelled north-east and entered the 'ladle' of Pei-tou".

The date of this event corresponds to AD 178 May 8. Hsüan-yüan lies in Leo and Lynx, but the actual star specified is uncertain. Pei-tou is the well known Plough, and the four stars forming the 'ladle' are α, β, γ, δ U. Maj. Several careful descriptions of the motions of comets which occurred during this same reign are given. The dates of these events are AD 178, 180, 182 and 188; in the latter year two separate comets are recorded. Translations of these observations are given by Ho Peng Yoke (1962) in his catalogue of comets and novae recorded in Far Eastern history.

One new star and five comets sighted in a period of 21 years is something of an achievement even by modern standards. However, planetary observations recorded in the Hou-han-shu are few and far between. There can be little doubt that only a minute proportion of the original observations of this type have survived. During Hsiao-ling's reign there are only five references to Venus, four to Mars and one to Jupiter; Saturn seems to have gone missing! (Mercury is, of course, very difficult to observe). Further, all but one of the planetary observations are of a very trivial nature, involving the passage of planets through various asterisms. Numerous essentially similar phenomena must have been noticed during this same period, but for one reason or another these have not found their way into the history. It might be remarked that this same feature, in varying degrees, is characteristic at all periods in Far Eastern history.

Despite the apparent dilatoriness of Hsiao-ling's astronomers in so far as planetary phenomena are concerned, each observation that

is reported was regarded as of very great omen value, and is
accompanied by a lengthy astrological commentary (post factum).
Were these few observations selected because the astrological
prognostications conformed to a basic pattern - foretelling the
downfall of the dynasty? But this is a subject on its own .....
(for a useful discussion see Bielenstein, 1950). The possibility
of falsification of astronomical records for astrological purposes cannot be ignored (see Chapter 2) and this might cast
doubt on the authenticity of the account of the guest star,
which so much concerns us here. With this in mind, we have felt
it advisable to check the reliability of the planetary data by
back calculation based on modern orbital elements. If the
observations prove to be reliable, then we can obtain some idea
of the accuracy which was achieved in astronomical observations.

The eight separate planetary observations which are recorded in
the astronomical treatise of the Hou-han-shu during the reign
of Hsiao-ling are as follows :
(a). "1st year of the Chien-ning reign period, 6th month,
Venus was in the west. It entered T'ai-wei and trespassed
against the star at the south end of the western 'wall'". The
date corresponds to AD 168 July 23 - August 20.
(b). "1st year of the Hsi-p'ing reign period, 10th month,
Mars entered within (chung) Nan-tou". Date : AD 172 November
4 - December 3.
(c). "2nd year (of the Hsi-p'ing reign period), 8th month,
(day) ping-yin, Venus trespassed against (fan) the front star
of Hsin". The year is equivalent to AD 173, but there was no
ping-yin day in the 8th month.
(d). "5th year (of the Kuang-ho reign period), 4th month, Mars
was within T'ai-wei and guarded (shou) P'ing". Date : AD 182
May 21 - June 18.
(e). "(5th year of the Kuang-ho reign period), 10th month,
Jupiter, Mars and Venus, all three, met at Hsü. They were
distant from one another 5 or 6 'inches' (ts'un) like a string of
beads". Date : AD 182 November 14.
(f). "3rd year (of the Chung-p'ing reign period), 4th month,
Mars retrograded and guarded the rear star of Hsin" . Date :
AD 186 May 7 - June 4.
(g). and (h). "6th year (of the Chung-p'ing reign period),
8th month, (day) ping-yin, Venus trespassed against the front
star of Hsin; (day) wu-ch'en, it trespassed against the middle
star of Hsin". Dates : AD 189 September 20 (g) and September
22 (h).

In the reduction of the above terse, but nevertheless fascinating
observations, we have used the planetary tables of Tuckerman
(1964) to interpolate the positions of the planets. We have
calculated the star positions from the data given by Newcomb
(1910). Let us consider each observation briefly in turn :
(a). From the star maps of Ho Peng Yoke (1962), the source of
most of our star identifications, the star at the south end of
the western wall of T'ai-wei is clearly σ Leo. Calculation
shows that on the evening of AD 168 July 23 - the 1st day of the
6th lunar month - Venus was only about 0.35 deg. from σ Leo.
and crossed into T'ai-wei. The observations is thus highly

accurate. We cannot say whether the day of the month was for-
gotten or lost at some later period.
(b).   The constituent stars of Nan-tou are the bright stars
μ, λ, φ, σ, τ, ξ Sag. During the 10th month Mars was much to
the west of Nan-tou, but in the 1st month of the following year
(i.e. 3 months later) the planet moved through the asterism,
passing between μ and λ Sag. at the beginning of the month.
The description 'within' Nan-tou would thus be accurate. Further
discussion is left until after treatment of the remaining
observations.
(c).   The stars of Hsin are well established as the bright stars
σ Sco. ("front star"), α Sco. ("middle star") and τ Sco. ("rear
star"). As stated above, there was no ping-yin date in the 8th
month so that we must leave the historical date in doubt.
However, calculation shows that on the evening of AD 173 September 19 - the day hsin-ch'ou in the 8th month - Venus was less
than 0.15 deg. from σ Sco. The month is thus correct, but for
some reason or other the day is completely in error.
(d).   During the first half of the 4th month, Mars was just
inside the well defined southern boundary of T'ai-wei (determined by the stars σ Leo., β Vir. and η Vir.), and passed less
than 5 deg. to the south of the asterism P'ing, whose constituents
are ξ, ν, π and ο Vir. The recorded description is thus accurate
enough, although the planet was not stationary, as use of the
term 'guarded' (shou) might imply.
(e).   The three planets Jupiter, Mars and Venus were several
degrees apart during the 10th month. However, continuing into
the 11th month, there was indeed a very close triple conjunction.
This is represented diagrammatically in Fig. 5.1, which shows the
relative positions (longitude and latitude) of the planets on
the evenings of AD 182 December 21 to 23. The configuration
on December 22 (the 10th day of the 11th month) agrees splendidly
with the graphic description of the text, "like a string of
beads", and furthermore the planets were within the lunar mansion
Hsü, as the record states.
(f).   Mars was certainly retrograding during the 4th lunar
month, but it did not approach close to τ Sco., the rear star
of Hsin, until the following month. The planet's stationary
point, reached in the middle of the 5th month, was less than
0.5 deg. to the west of τ Sco., but about 3 deg. to the north of
it, so that for practically the whole of the 5th month Mars
would appear to be suspended above the star, and could thus
accurately be described as "guarding" it.
(g). and (h).   On the evening of AD 189 September 20, Venus was
less than 0.35 deg. from σ Sco., Two days later, the planet was
only about 0.20 deg. from α Sco. The stated dates are thus
correct and both observations are particularly accurate.

From the above analysis, it is clear that there is no evidence
whatever of deliberate fabrication of astronomical observations
during Hsiao-ling's reign. Numerous celestial events may have
been omitted - deliberately or otherwise - but those which are
recorded are of high reliability, and allow us to place considerable confidence in the account of the guest star of AD 185.
The planetary records fall into two categories - those for which
both the month and day are stated, and those for which only

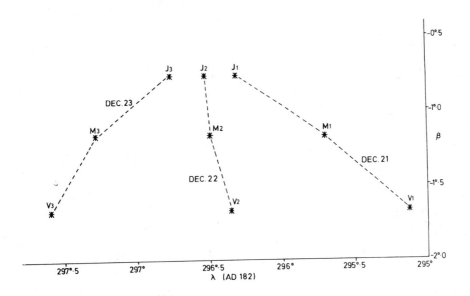

Fig. 5.1. Triple conjunction of Jupiter, Mars
and Venus AD 182 December 21-23. Positions
shown are for 1 hour after sunset.

the month is given. Of the former we have found two instances
where the date is exact and a third where the month is correct
but the day is completely erroneous. Where only the month is
given, this is more often than not incorrect, but normally no
more than a month in error. If errors of this amount can occur,
it seems reasonable to assume that in the case of the observation
listed as (b) above, the suggested error of 3 months is real.

Returning to the new star of AD 185, there is no record of it
elsewhere in the history (Harvard - Yenching Index No. 41, 1966).
However, as Fan-yeh, the compiler of the rest of the Hou-han-shu,
wrote long after Szŭ-ma Piao, it is possible that the latter had
access to material which was not available to Fan-yeh. The date
of first appearance of the star, which is given to the day, is
probably accurate; as noted earlier this corresponds to AD 185
December 7. However, only the month of disappearance is stated,
and, as we have just seen, this should be treated with caution.

The duration of visibility has been variously interpreted.
Whether it is approximately 8 or 20 months turns on the interpretation of the term hou-nien in the text. At the beginning
of this chapter we have translated this as 'the year after next',
implying the longer duration and a date of disappearance between
AD 187 July 24 and August 21. However, Ho Peng Yoke (1962)
renders the expression as 'the following year' and gives the date
when the star ceased to be visible as AD 186 July 5 to August 2.

Hsi Tsê-tsung (1955) originally preferred this latter interpretation, but in his later paper in collaboration with Po Shu-jen
(1965) he favours the longer duration. The period of visibility
of the star is so important that it is worth considering this
seemingly minor technicality in some depth.

In modern Chinese usage the expression hou-nien always means
'the year after next', just as hou-jih means 'the day after
tomorrow'. The equivalents of our 'next year' and 'tomorrow' are
respectively ming-nien and ming-jih. On the early use of hou-
nien the dictionary of Morohashi (1955) is not too definite, and
in order to ascertain whether the meaning in later Han times
(or at least when the astronomical treatise of the Hou-han-shu
was written) was the same as at present we did a count of the
relative frequency of occurrence of ming-nien and hou-nien in
the remainder of the astronomical treatise. There are twelve
examples of ming-nien, but only this single isolated example of
hou-nien. Frequently hou-x-nien (where x is a small number) is
also used, meaning x years later, but the usage of hou-nien
itself is unique, suggesting that it was deliberate. We thus feel
that we can adopt the longer duration with a high degree of confidence.

A period of visibility as long as some 20 months makes the probability that the star was a supernova extremely high; a comet
can be entirely ruled out. It should be pointed out that a duration of more than a year implies that the text remains silent
on the disappearance and recovery of the star around the time of
conjunction with the Sun (heliacal setting and rising). However,
this is not a major difficulty. The supernova of AD 1054 is
far better documented than the star under consideration (see
Chapter 8 below), but although both the dates of first sighting
and final fading are given, corresponding to a period of visibility of nearly two years, there is not a single reference to
heliacal setting or rising.

The asterism Nan-mên, in which the guest star appeared, was
probably the southernmost star group visible from central
China. In the astronomical treatise of the Chin-shu we read
"The two stars of Nan-mên ("Southern Gate"), situated south of
K'u-lou, form the outer gate of the heavens and govern garrison
troops" (trans. Ho Peng Yoke, 1966). The importance of this
description lies in its assertion that Nan-mên consisted only of
two stars. Ho Peng Yoke in his commentary on the above quotation was of the opinion that the two components were α and ε Cen.,
but in his star maps at the end of his book he links β and ε
Cen. Hill (1967) in his pioneering discussion of the possible
radio remnants of the new star of AD 185 preferred to leave the
question open, merely stating that "Nan-mên was apparently
associated with the stars α, β and ε Cen". However, the triangle
formed by these three stars covers a fairly large area of sky
(some 10 square degrees). In order to clarify the issue,
Stephenson (1975) undertook a careful examination of copies of
oriental star maps. The originals of these vary in date from
about AD 900 to 1750. This investigation has since been extended
to cover the very early (around 2,000 years old) star maps in the

Hsing-ching ("Star Manual").   All of these charts are agreed that
Nan-mên consisted only of two stars, and although most are crudely
drawn, estimates of the right ascension and declination of the
asterism (especially from the later maps) are sufficient to prove
that we must indeed choose from α (mag. -0.3), β (mag. +0.6) and
  ε Cen. (mag. +2.3).   There are no other bright stars in the
vicinity (all others are fainter than mag. +4), and, as the
asterism was only a few degrees above the horizon in Central China,
no other nearby stars would be expected to be readily visible.

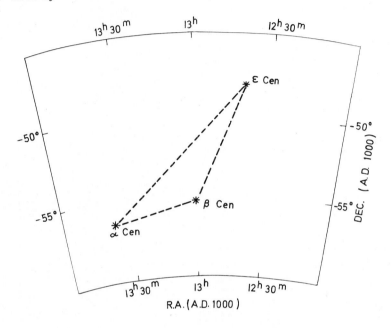

Fig. 5.2. Positions of α, β and ε Cen. at epoch AD 1000.

The celestial co-ordinates of α, β and ε Cen. are shown
diagrammatically in Fig. 5.2 for AD 1,000 - a typical mean epoch
for comparison with the representation of the asterism on the
various oriental star maps.   The figure should be compared with
Fig. 5.3.   This is an illustration of a star map in the Ku-chin-
t'u-shu-chi-chêng, a Chinese encyclopedia dated from AD 1725, but
containing star charts (pre - Jesuit) from more than a century
earlier.   The region of sky covered is roughly from about 10^h to
14^h right ascension and from -60° to + 50° declination.   This
chart is almost unique among oriental representations in that it
makes some attempt to distinguish between stars of different
brightness.   Three classes are recognised.   Undoubtedly the
three bright stars are α Vir. (near the centre of the chart) and
α and β Cen. near the lower edge.   It is evident that the map is
not drawn to scale (a common failing), but nevertheless there can
be no question that Nan-mên was understood to consist only of α

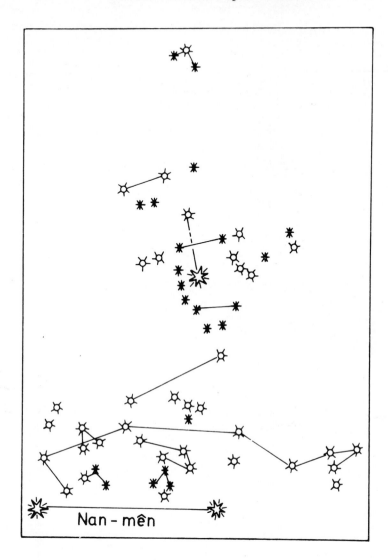

Fig. 5.3. A copy of an early 17th century star chart showing a small section of the sky. We have omitted the Chinese characters identifying the various asterisms and inserted the romanisation for Nan-mên, where SN 185 appeared. The three bright stars are α Vir. (near the centre of the map) and α and β Cen., forming Nan-mên.

and β Cen. (there is an isolated star further north which could possibly represent ε Cen.).

The other oriental star maps which we have consulted do not differentiate between stars of different brightness, cf. Plates 3 and 4.   In order to determine the components of Nan-mên, we found the most satisfactory technique was to measure on each chart the angle between the line joining the two stars and an estimated circle of declination passing between them.   This angle was in all cases between 0° and $+35^O$ (measured towards the east), with a mean of $+15^O$.   In Fig. 5.3 it is close to 0°, but in Plate 4 it is about $+30^O$.   This is in acceptable agreement with the actual angle for α and β Cen $(+10^O)$, but incompatible with the very large angles for α and ε Cen or β and ε Cen.   Accordingly we have no hesitation in identifying α and β Cen as the two stars of Nan-mên.   This considerably reduces the area of search.

Let us now consider the position of the new star in relation to Nan-mên.   It was described as "within" (chung) the asterism. The usual expression to denote the location of a planet or new star is the rather vague term yü, which may be translated "at" or "in the vicinity of".   However, chung, which is comparatively rare. is more specific.   Depending on the context, it can be rendered as "middle" or "within".   In its former capacity it is used in the name for China itself - Chung-kuo ("The Middle King-dom").   There are various examples of the latter meaning in the astronomical treatise of the Hou-han-shu.   Thus we find several instances in Chapters 21 and 22 where a planet is said to have entered within an asterism.   An occultation of Venus in AD 72 is described as "Venus entered within the Moon" (Chapter 21). Sunspots in AD 188 and 189 were identified as follows: (AD 188) "a black vapour as large as a melon was within the Sun"; (AD 189) "the Sun was orange in colour and within it there was a black vapour like a flying magpie.   After several months it melted away".

In order to check the precise significance of chung where an asterism is concerned, we examined the various planetary observations recorded in the Hou-han-shu.   In most cases the planets were said to enter within star formations containing a number of rather faint stars, making their identification questionable.   However, two observations, both involving the planet Venus seemed very reliable since the star group concerned (the same in each case) was bright and the location within it was given.   In both AD 117 and 125 Venus is reported as entering within the "mouth" (k'ou) of Nan-tou.   This well-defined lunar mansion in Sagittarius, whose name may be translated as "Southern Dipper" is very similar in form to the better known "Northern Dipper" or "Plough" in Ursa Major.   The constituent stars are readily identified as φ , λ , μ , σ , τ and η Sgr. for there are no other bright stars in the vicinity.

The dates of these two Venus observations, which are recorded in chapter 21 of the Hou-han-shu, correspond to AD 117 October 23 and AD 125 October 24.   It is perhaps noteworthy that these dates are almost exactly 8 years apart (i.e. 5 synodic periods of Venus)

but it is doubtful if the Chinese astronomers were aware of the
Venus cycle of this length.   Fig. 5.4 shows the positions of the
stars of <u>Nan-tou</u> at the mean epoch AD 121, corrected for precession and proper motion.   The two locations of Venus are calculated
for the evenings of AD 117 October 23 (V1) and AD 125 October 24
(V2), and the approximate motion of the planet in one day is
shown for comparison.   There can be no doubt that the "mouth" of
<u>Nan-tou</u> is formed by φ, σ, τ and ε Sgr.   From Fig. 5.4, as well
as the evidence that both recorded dates are exactly correct, it
is clear that the observations are accurate to within a small
fraction of a degree.   On each occasion Venus was just entering
the "mouth" of the asterism.   <u>Chung</u> was indeed a precise term.

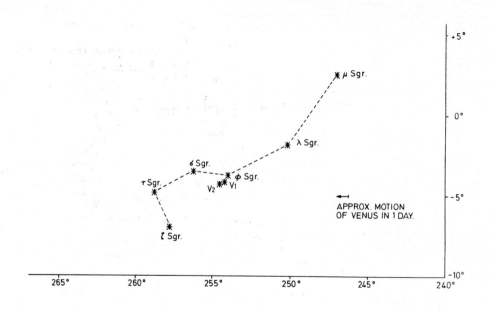

Fig. 5.4. Observations of Venus entering within
the 'mouth' of the asterism <u>Nan-tou</u>, AD 117
October 23 (V1) and AD 125 October 24 (V2).
Positions shown are for 1 hour after sunset.

In the context of the guest star record, we can now place the star
roughly between α and β Cen., and in looking for its remnant
consider only those SNRs which answer this location.

The final observational question concerns the apparent brightness
of the supernova.   It was never far above the horizon at Lo-yang
(the meridian altitudes of α and β Cen. were respectively only 3°

Fig. 5.5. A simplified map of the Lo-yang area.

and 5°) and would only be above the horizon for a very few hours
each day. We have consulted a detailed (1 : 250,000) relief
map of the Lo-yang area with contours at 100 m intervals. This
is shown in simplified form in Fig. 5.5, with the site of the
ancient capital indicated; to avoid unnecessary confusion only
the 500 m contour is included. An horizon profile is shown as
Fig. 5.6, and this indicates that the southern horizon from
Lo-yang was relatively, unobscured (more later). At low elevation atmospheric absorption would be considerable, amounting to
some 3 mag. Probably the optical dispersion and apparent physical
size of the object (as described in the text) can be accounted
for in this way. Because of the poor visibility of the star we
must assume an error in the month of final disappearance. Even
at the beginning of the 6th lunar month in AD 187 (late July)
the star was setting almost simultaneously with the Sun, so that
it would only be above the horizon during the hours of daylight.
However, a month previously there would probably be a brief
opportunity to observe it in a dark sky after sunset. (N.B. This
difficulty would still arise even if we preferred a duration of
less than a year). Several instances of an error of one month
have already been encountered - the planetary observations
discussed in this chapter - and it seems legitimate to suppose
this to be the case here, although, of course, we can never
prove this. We thus make the tentative suggestion that the star
disappeared in the 5th lunar month, corresponding to AD 187 June
24 to July 23.

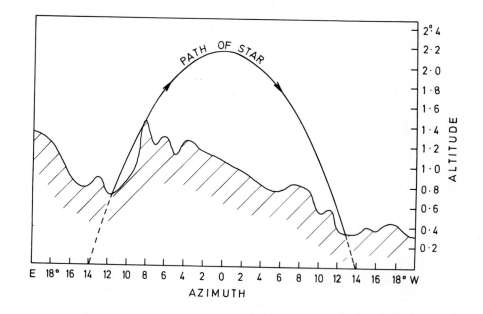

Fig. 5.6. Profile of the southern horizon from
Lo-yang showing the apparent diurnal path of a
star having the same location as G315.4-2.3.

In order to obtain some preliminary estimate of the apparent
magnitude of the star at maximum we can use the method outlined
in Chapter 1. This makes use of the duration and assumes typical
Type I supernova parameters. Taking the final magnitude as +2
(allowing about 3 mag. for atmospheric extinction) gives a rough
estimate for the magnitude at maximum of about -8. The extraordinary brilliance is highlighted by the fact that, if the
stated date of appearance of the guest star is correct, its
visibility must have been seriously impaired by the dawn-glow,
since the star was rising just ahead of the Sun (more later).
A distance in the region of 1 to 2 kpc thus seems to be indicated.
We therefore seek a nearby remnant of age 1800 years lying roughly
between α and β Cen. Clark and Caswell's (1976) catalogue of
SNRs lists four remnants in this area of Centaurus ; namely
G311.5-0.3, G315.4-2.3, G316.3-0.0, and G315.4-0.3. This region
has now been particularly well surveyed, and there are no other
sources which could be considered as possible SNRs. Fig. 5.7
shows the position of the four candidate SNRs with α and β Cen.
for epoch AD 185. (Proper motions of the SNRs have been ignored;
however this is unlikely to be the cause of significant error.)
As noted in Table 4.2, the only SNR in this region that appears
to be near enough and of small-enough diameter (i.e. young
enough) to be the remnant of AD 185 is G315.4-2.3. Nevertheless,
to confirm this initial assessment we will consider here all
four SNRs of Fig. 5.7.

G311.5-0.3 was first identified as an SNR by Shaver and Goss
(1970) from their partial southern Galactic Survey, and it was
included as Number 35 in the early SNR catalogue of Downes (1971).
The source is of small angular diameter and is confused with an
adjacent H II region (G311.5-0.5). Presently available radio

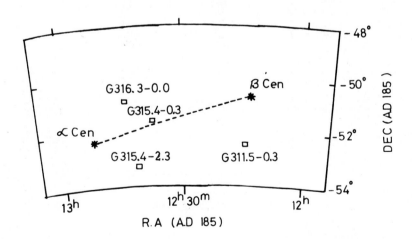

Fig. 5.7. The position of the four candidate
SNRs for the supernova of AD 185, with α and β Cen.

maps have insufficient resolution to detect peripheral brightening.   The H I absorption technique described in Chapter 4 for
distance determination gives a minimum distance estimate of 6.6
kpc (Caswell et al, 1976);   in fact the small angular diameter
and a low radio-frequency flux density measured for the source
suggest that it is probably on the remote side of the Galaxy.
The large inferred distance for G311.5-0.3 makes it a <u>highly</u>
<u>unlikely</u> candidate for the remnant of the supernova of AD 185.

G316.3-0.0 was catalogued as MSH 14-<u>57</u> in the pioneering 85 MHz
southern radio survey of Mills, Slee, and Hill (1957), and was
included in the Milne (1970) catalogue of SNRs.   High-resolution
maps show the double-lobe configuration characteristic of an SNR
viewed across the direction of the interstellar magnetic field
(see Chapter 4.)   A previously unpublished radio map of the source
made by the authors with the Molonglo radiotelescope at 408 MHz
is shown as Fig. 5.8.   An H I absorption distance estimate
(Caswell, 1967) places this SNR at a distance of ⩾7.2 kpc with
diameter ⩾36 pc.   This extreme distance estimate must preclude
G316.3-0.0 from consideration as the remnant of the supernova
of AD 185.

Fig. 5.8. A 408 MHz map of the SNR G316.3-0.0.

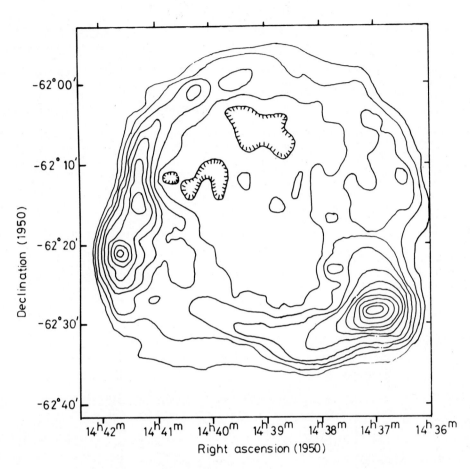

Fig. 5.9. A 408 MHz map of the SNR G315.4-2.3, the likely remnant of the supernova of AD 185.

The radio source G315.4-2.3 was also present in the 85 MHz catalogue, as MSH 14-63. Hill (1964) was the first to present evidence of its possible supernova character, and he later published the results of observations at 1410 and 2650 MHz clearly indicating a non-thermal spectrum and peripheral brightening (Hill, 1967). In Fig. 5.9 we show the highest resolution radio map presently available compiled from observations made by the authors with the Molonglo radio telescope at 408 MHz. The source is 40 arc min. in angular diameter, and shows a well-defined broken shell structure. Comparable resolution observations at 5000 MHz made with the Parkes radiotelescope (Caswell, Clark and Crawford, 1975) show a similar structure, and the 408 MHz and 5000 MHz flux density estimates yield a non-thermal spectral index of -0.62. Milne (1972) detected significant

polarization from the regions of peak radio brightness, commen-
surate with the non-thermal nature of the source.   Optically the
remnant consists of a set of bright filaments, designated as
RCW 86.   A weak elongated filament lies along the northern edge
of the radio shell, and a set of bright filaments (shown as
Plate 7) coincides with the region of enhanced radio emission in
the south-west corner of the remnant.   A concentration of early
B stars at a distance of 2.5 kpc was the basis of Westerlund's
(1969) suggestion of this being the possible distance to RCW 86.
As already noted, a somewhat smaller distance is preferred for
the remnant of the supernova of AD 185 because of its probable
extreme brightness at maximum.   This possible distance discrepancy is of little consequence in the consideration of G315.4-2.3
as the remnant of the AD 185 supernova, since both distance estimates are highly subjective.   Nevertheless, on the basis of the
brightness of the optical filaments alone, the relative proximity
of the remnant to the Sun is beyond dispute, making G315.4-2.3
a highly likely candidate for the AD 185 supernova.   In addition,
G315.4-2.3 is the southernmost of the four SNRs in the region,
with meridian altitude at Lo-yang about 2.2° (allowing for
refraction).   A star this close to the horizon would be likely
to suffer the severe dispersion ("it showed the five colours")
and distortion ("it was as large as half a mat") described in the
text for the guest star.

Fig. 5.10. A 408 MHz map of the SNR G315.4-0.3.

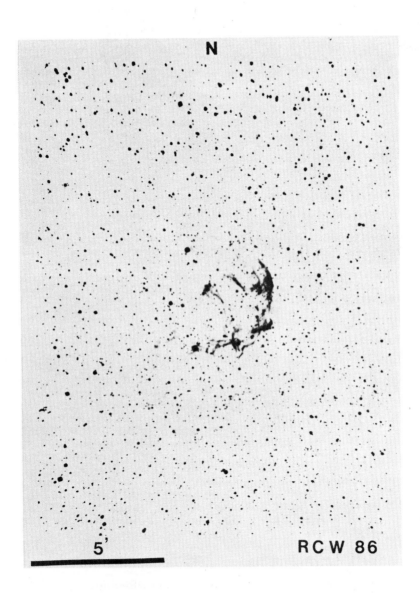

Plate 7. The optical remnant RCW 86.

We come now to the final SNR to be considered, G315.4-0.3.
This is one of 28 southern radio sources recently recognised to
be SNRs by Clark, Caswell and Green (1975). The Molonglo 408 MHz
map for G315.4-0.3 is shown in Fig. 5.10. The source is ≃16 arc
min angular diameter, displaying an enhanced emission in the south-
west quadrant. The point source on the north-eastern boundary is
believed to be extragalactic, and was extracted in calculating a
non-thermal spectral index of -0.47 for the remainder. There
is no evidence of associated optical filamentary structure. No
independent distance estimate is available for the remnant.
However the "Σ-D" approach described in Chapter 4 suggests
a distance of about 8 kpc for G315.4-0.3, but with large uncertainty since it may possibly be a nearby object with anomalously
low surface brightness. (RCW 103 is recognised to be such an
SNR, with a well-determined H I absorption distance of 3.3 kpc,
but a Σ-D distance estimate of 8.7 kpc).

To summarize : of the four candidate SNRs shown in Fig. 5.7,
G311.5-0.3 and G316.3-0.0 must be rejected on the basis of their
large distances. Although G315.4-0.3 fits perfectly the Chinese
description of the guest star of AD 185 lying 'between' (chung)
α and β Cen., its distance inferred from a Σ-D approach is
much too high ; an H I absorption distance estimate is needed
to investigate the possibility that the remnant may be closer.
In the absence of such data, the available evidence continues
to support the suggestion that G315.4-2.3 (RCW 86) is the probable
remnant of AD 185.

As mentioned earlier, a star in the position of G315.4-2.3
would be only 2.2° above the horizon at Lo-yang when on the
meridian. The apparent diurnal path of a star at this location
is depicted in Fig. 5.5. Under normal circumstances, a bright
star would be readily visible at this altitude, although atmospheric absorption would be heavy - as much as 4 magnitudes
(at very low altitudes light loss in the atmosphere increases
rapidly towards the horizon). However if G315.4-2.3 is in fact
the remnant, the star at discovery would be to all intents and
purposes a daylight object. Before sunrise the star would be
scarcely visible (altitude at sunrise only 1.7°), and the best
opportunity to view it would be after the Sun had risen. Assuming a required magnitude of -4 for visibility in daylight leads to
a magnitude of at least -8 for the new star. This extreme brightness is in agreement with the preliminary estimate made earlier
on the basis of period of visibility and assuming typical Type I
supernova parameters.

Our preferences for a 20 month duration of visibility, an extreme
brightness at maximum (≃ mag. -8), and G315.4-2.3 as the remnant,
do in fact allow an interpretation of the supernova of AD 185 as
either Type I or Type II. If Type I (assuming visual absolute
magnitude at maximum -19), allowing an interstellar absorption of
1 mag. / kpc gives a source distance of ≃ 1 kpc, a remnant diameter of ≃12 kpc, and an average expansion velocity of ≃6,500
km/sec. The remnant would be at greater distance if the estimate
of maximum apparent magnitude is high, although a distance as
great as 2 kpc would be possible only if one were to assume an

apparent magnitude of -5; as noted above such a value appears
to be incompatible with the supernova's inferred day-time sighting
at low meridian altitude. If Type II (assuming visual absolute
magnitude at maximum -17.5), the corresponding parameters are
distance ≈600 pc, diameter ≈7 pc, and average expansion velocity
≈4,000 km/sec. In principle either set of parameters would be
acceptable, and there is clearly nothing in the historical record
which would enable us to distinguish between Type I and Type II.
Nevertheless, and this judgement must be conjectural in the absence
of convincing experimental evidence of the behaviour of Type II
light curves beyond about 150 days, the long period of visibility
tends to favour a type I interpretation.

Since the guest star of AD 185 is the oldest supernova recorded
historically, we have felt justified in subjecting the single
text referring to it to close scrutiny. In particular, it was
believed necessary to test the reliability of the positional
description and the recorded duration of visibility, since an
unambiguous identification of the remnant of the supernova
depended on this information. Despite minor errors in the dates
recorded for certain astronomical events, there can be no doubting
the essential reliability of the Chinese observations at this time.
Their single record of the supernova of AD 185, and the identification of its probable remnant based on the historical description,
will prove invaluable in investigating the evolution of supernova
remnants up to an age of 2000 years.

Chapter 6

THE CHIN DYNASTY GUEST STARS

Continuing in chronological order, the next two new stars which rate as possible supernovae were observed in AD 386 and 393. As both stars were recorded only by the Chinese, and occurred in the same reign period, it seems appropriate to consider them together, especially as each description is very brief.

Both stars appeared during the latter years of the Chin Dynasty in China, but the earliest records of them which we possess are contained in a section of the astronomical treatise (Chapter 25) of the Sung-shu. The entire history is the work of one man, Shen Yüeh, who lived between AD 441 and 513. The astronomical treatise contains many observations dating from the Chin period.

After Shen Yüeh completed the Sung-shu, more than a century was to elapse before the compilation of the official history of the Chin Dynasty - the Chin-shu. From Ho Peng Yoke (1966), this latter work was compiled by a team of scholars under the edict and supervision of Emperor T'ai-tsung, the second ruler of the T'ang Dynasty. At that time there existed no less than 18 different versions of the history, but all of these were found to be inadequate to serve as an official history. The astronomical chapters of the Chin-shu were in all probability written by Li Shun-fêng (AD 602 - 670), an eminent astronomer and mathematician, but because he wrote so long after Shen Yüeh, it is likely that the latter had access to more direct sources than Li Shun-fêng.

The astronomical treatise of the Sung-shu follows the pattern of that in the Hou-han-shu : The various types of observation are not classified according to type. However, in the corresponding section of the Chin-shu, astronomical records are classified under as many as nine different headings. The new stars of AD 386 and 393 are to be found in the section entitled "Ominous stars and guest stars", which forms part of chapter 13 of the history.

Ho Peng Yoke's (1966) excellent translation into English of the astronomical chapters of the Chin-shu has made this work generally accessible in the West. No such translation of the Sung-shu exists, although in his commentary on the Chin-shu records Ho Peng Yoke frequently quotes from it.

The reports of the two new stars in both the Sung-shu and the Chin-shu are very much alike, both in the description of the stars and in the accompanying astrological commentary. We have felt it desirable to give translations of the accounts of the stars from both sources, but to avoid unnecessary repetition have quoted the astrological interpretations from the later

Chin-shu only.

AD 386 :
Sung-shu (Chapter 25). "11th year of the T'ai-yüan reign-period,
3rd month, there was a guest star at Nan-tou until the 6th month
when it was extinguished (mieh)".
Chin-shu (Chapter 13). "11th year of the T'ai-yüan reign-period,
3rd month, there was a guest star at Nan-tou until the 6th month
when it vanished (mo)".

The months of appearance and disappearance correspond to AD 386
April 15 - May 14 and July 13 - August 10.
Astrological Prognostication. "The interpretation was military
activities and an amnesty. After this (the people of) Szǔ, Yung,
Yen, and Chi were frequently called up for military service.
Later in the 12th year, 1st month, there was a general amnesty,
and in the 8th month another general amnesty".

AD 383 :
Sung-shu (Chapter 25). "18th year of the T'ai-yüan reign-period,
2nd month, there was a guest star within (chung) Wei until the
9th month when it was extinguished (mieh)".
Chin-shu (Chapter 13). "18th year (of the T'ai-yüan reign-period),
2nd month, there was a guest star within (chung) Wei until the
9th month when it vanished (mo)".

The months of appearance and disappearance correspond to AD 393
February 27 - March 28 and October 22 - November 19. (It should
be noted that there was an intercalary 7th month in the Chinese
calendar during this interval).
Astrological Prognostication. "The interpretation was military
activities and deaths at Yen. In the 20th year Mu-jung Ch'ui
and Hsi-pao attacked Wei, but were defeated and more than 10,000
men were killed. In the 21st year (Mu-jung) Ch'ui died, and
eventually his state was ruined".

The great similarity between the records of AD 386 and AD 393 is
self-evident, suggesting that the original reports were reduced
to a stereo-type pattern, presumably by Shen Yüeh. The Chin-shu
copies the style of the Sung-shu, changing only the final
character (mieh to mo), but without altering the meaning in any
way.

The imperial annals of the Chin-shu (Chapter 9) are silent on
the new stars of AD 386 and 393 but in any case they contain
few astronomical observations - mainly solar eclipses. We do
not expect any additional records from the Sung-shu since the
annals of this history only deal with the Liu-sung Dynasty itself.
We have searched several of the early Chin histories for references to the guest stars, but without success.

It is perhaps something of a coincidence that like their counterpart in AD 185, the stars of AD 386 and 393 heralded the downfall
of a dynasty. The Chin Dynasty was far weaker than the Han, and
as early as AD 317 the western half of the empire was lost to
barbarians. The capital was then removed to Chien-k'ang (modern

Nan-ching), where it remained until the fall of the dynasty in
AD 420. Hsiao-wu, who was emperor at the time when the guest
stars appeared, reigned from AD 373 to 397. He was a weak,
incompetent ruler who did nothing to stem the tide of rebellion.
After his death at the age of 34 (he was murdered while hopelessly
drunk by his favourite concubine) the dynasty dragged on for a
further 23 years before its final end. The astrological prognostications accompanying the various astronomical observations
made during Hsiao-wu's reign reflect the troubled state of these
times and this is particularly true of the guest star records.
Two planetary observations at this time were taken as portents
of the emperor's death - an occultation of Mars by the moon
(AD 393 February 20), and a conjunction of Jupiter with two small
asterisms (AD 396 July 22 - August 19).

It is unfortunate that the records of both new stars tell us so
little of astronomical interest. No reference to brightness or
colour is made and only the months of discovery and last sighting
are given. Bearing in mind the political turmoil of these times,
the complete fabrication of astronomical records and astrological
prognostications under political pressures might seem to have
been possible. Therefore, it seems desirable once again to
test the reliability of astronomical records in general around
this period, and in particular to check the accuracy of dates
recorded only to the nearest month.

During the reign of Emperor Hsiao-wu, six observations relating
to the motions of planets are reported for which the date is
given only to the nearest month. All observations were made
during the second part of his reign - the T'ai-yüan reign-period
- in which the guest stars appeared. In the following
translations of the observations the year is given in terms of
the T'ai-yüan reign period, the first year of which corresponds
to AD 376-377. The text used is the Sung-shu, but any significant discordances between this and the Chin-shu will be commented
upon.
(a). "(1st year) 9th month, Mars trespassed against K'u-hsing
and Ch'i-hsing and then entered Yü-lin". Date AD 376 September
30 - October 28.
(b). "3rd year, 6th month, Mars guarded Yü-lin". Date AD 378
July 11 to August 9. The Chin-shu gives the date as "2nd year,
2nd month", which corresponds to AD 377 February 24 to March 25.
(c). "(13th year, 12th month) Mars was at Chüeh and K'ang".
Date AD 389 January 13 to February 11.
(d). "14th year, 12th month, Mars entered Yü-lin". Date AD 390
January 3 to 31.
(e). "19th year, 10th month, Venus, Saturn, Mars and Mercury
met at Ti". Date AD 394 November 10 to December 8. This
observation is reported in the Chin-shu but not in the Sung-shu.
(f). "20th year, 6th month, Mars entered T'ien-chün". Date
AD 395 July 4 to August 1.

It will be noticed that each of the observations mentions the
planet Mars, but there seems to be no special reason for this.
There are plenty of sightings of Jupiter and Saturn during this
same period, but here the precise date is given (this also

applies to a number of further observations of Mars).

A brief discussion of each record is as follows :
(a). K'u-hsing, Ch'i-hsing and Yü-lin are all rather faint
asterisms lying mainly in Aquarius. During the 9th month Mars
was moving eastwards so that the first asterism reached would be
K'u-hsing. This consists of two stars, of which the westernmost
is µ Cap. As it happens, the planet did not make its closest
approach to µ Cap. until the 3rd day of the 10th month. In this
latter month Mars passed close to Ch'i-hsing and skirted the northern edge of Yü-lin. We have here an error of a month in the
recorded date, but otherwise the motion of the planet was accurately described.
(b). As already pointed out, there is considerable discord
between the dates recorded in the Sung-shu and the Chin-shu
for this event. In reality, neither date is correct : during
the two years in question, Mars was only in the vicinity of
Yü-lin during the 3rd and 4th months of the 3rd year. There is
an unaccountable error here.
(c). Chüeh, which consists only of the two bright stars α and
γ Vir., is a particularly well-defined asterism, and K'ang, also
in Virgo and lying some 10 deg. to the east of Chüeh, is also
fairly bright. About the 20th day of the 11th month, Mars moving
eastward, passed between the stars of Chüeh, but during the 12th
month the planet was almost stationary between the two asterisms.
The recorded month is thus probably correct.
(d). Yü-lin is an extensive asterism consisting largely of
faint stars. It lies a little to the south of the ecliptic in
Aquarius. During the latter half of the 11th month and the first
day or two of the 12th month, Mars, moving rapidly eastward,
passed along the northern edge of Yü-lin. In view of the
difficulty in identifying several of the constituents of this
asterism, it is not possible to say just when the planet actually
entered the asterism (if at all). The recorded date thus may be
correct or a month in error.
(e). This is perhaps the most interesting observation of the six.
Four planets in close proximity to one another is a rather rare
occurrence. Calculation shows that from the 15th to the 18th
day of the 10th month (November 24 to 27) the planets Mercury,
Venus, Mars and Saturn covered a range of no more than 14 deg.
What is particularly important is that on the first two of these
dates all four planets were within the lunar mansion Ti, in
accord with the Chinese description. Here the recorded month is
unquestionably correct.
(f). T'ien-chün, in Cetus, is a fairly well defined asterism
lying a little to the south of the ecliptic. Commencing about
the 20th day of the 6th month and continuing into the beginning
of the following month, Mars moved along the northern edge of
the asterism, but without actually entering it. As the term
chung is not used in the text, the description may not be too
critical. Here the date is probably correct.

Once more, we have a vindication of the reliability of ancient
Chinese astronomical observations. It does seem that where only
the month is recorded, errors of a month in the actual date are
quite frequent, but this is the normal limit. It is significant

that the one example of a major error occurs in (b), for which
there is marked discord between the two historical sources.

In view of the above remarks, we should regard the agreement in
date between the two separate records of each star as confirmation
of reliability. However, whereas an error of a month in the first
or last sighting of the AD 393 object would not alter the fact
that the duration was unusually long, it would seriously affect
the length of visibility of the AD 386 star (reported to be 3
months). The earlier guest star is thus of very uncertain nature,
but the likelihood of the later star representing a supernova
outburst seems fairly high.

Let us now consider what can be deduced about the position and
brightness of each object in turn :
AD 386
Nan-tou, in or near which the guest star appeared, has already
been discussed in Chapter 5 under the Venus observations. It is
shown in Fig. 6.1 for epoch AD 386. The position of the new
star in relation to Nan-tou is quite vague, for yu ("at" or "in
the vicinity of") is a very general expression. It is interesting to note that the star appeared near one of the two points
where the galactic equator cuts the ecliptic. None of the planets
was in the vicinity while the star was visible, but it is a
pity that there are no references to the Moon passing close to
the star, because this might have helped to fix its position
better. In AD 1605, Venus approached to within half a degree
of Kepler's supernova, and this was duly recorded by the Korean
astronomers (see Chapter 11). In the present case we must
consider any young SNR near Nan-tou as a possible remnant.
In fact there are only three remnants in the vicinity which
could possibly correspond, and these are discussed below.

The brightness of the star is not estimated in the record, but
we can make some useful inferences. The star was almost in
opposition to the Sun throughout the period of visibility so its
disappearance was not caused by heliacal setting. From the
discussion in Chapter 3 of the almost complete lack of Far
Eastern sightings of Mira Ceti, an apparent magnitude of +1.5
or brighter seems needed to make detection probable. Atmospheric
absorption can be neglected since the meridian altitude (about
35 degrees) was fairly high. A supernova of either Type I or
Type II having this apparent brightness at maximum would remain
visible to the unaided eye for about 3 months. If the recorded
duration is correct, a reasonable estimate of the magnitude
would be near +1.5 (with some uncertainty). Following the usual
practice, a distance of between 5 and 10 kpc seems probable, but
the combination of poor positioning and uncertain duration make
this star a rather weak supernova candidate.

From Table 4.2 only the remnant G11.2-0.3, lying a few degrees
west of Nan-tou for epoch AD 386, has a distance estimate
($\geqslant 5$ kpc) within the above range. The H I absorption distance
estimate corresponds to a linear diameter of the source of
$\geqslant 6$ pc; the lower limit value would be unacceptably small,
and even if the source were at 10 kpc (with diameter $\simeq$ 12pc),

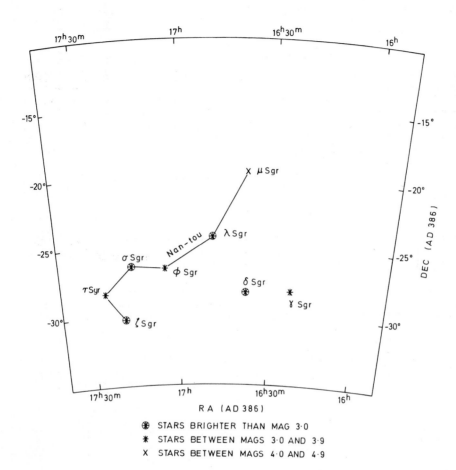

Fig. 6.1. The asterism <u>Nan-tou</u>.

it must have undergone a comparatively slow expansion if it was
in fact the remnant of AD 386.

Only two other remnants fit the vague positional description
of the guest star;   these are G11.4-0.1 and G12.0-0.1, both
recently discovered by Clark, Caswell, and Green (1973, 1975).
Each remnant is of small angular diameter and low surface brightness, suggestive of extreme distance;   in fact the Σ - D distance estimates are 15.9 and 19.7 kpc respectively. Despite the
possibility that one of them may be of anomalously low surface
brightness, and at closer distance, neither stands out as the
likely remnant of the AD 386 event.

Because of the short recorded duration for the guest star of
AD 386 and the absence of a completely acceptable remnant,
this event can at best merely be described as a <u>possible</u> supernova.

AD 393

Here we have a much more satisfactory object. The duration is
considerably longer and the position rather carefully described.
The star was said to be "within" (<u>chung</u>) the asterism <u>Wei</u>.
We have already discussed the use of <u>chung</u> (Chapter 5) and found
it to be very precise. Additionally, <u>Wei</u>, the "Tail of the
Dragon" in oriental astronomy and the "Tail of the Scorpion" in
the west, is a very well defined asterism. Fig. 6.2 is drawn for
the epoch AD 393, precession and proper motion having been
allowed for. All stars within the boundary of the figure having
an apparent magnitude brighter than +5 are represented. It
requires little imagination to trace the outline of Wei.

Any star described as "within" <u>Wei</u> would have to lie inside the
polygon completed by joining ν <u>and</u> ε Sco. This region is almost
bisected by the galactic equator near which the majority of
SNRs lie. It is only a pity that the area enclosed is rather
large - some 40 square degrees.

Once more we have no direct reference to the brightness of the
star. The recorded period of visibility was close to 8 months,
but this does not allow for the possibility of an error of a

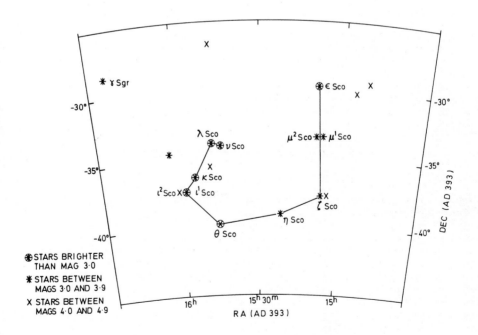

Fig. 6.2. The asterism <u>Wei</u>.

month in the date of first or last sighting. As it happens, the recorded date of last visibility is almost certainly in error. Even as early as the 1st day of the 9th lunar month (October 22) the star was already setting in a light sky only about 15 minutes after the Sun (from the latitude of Nan-ching, 32° N). To be still visible under such conditions would require a bright object, so that 8 months before, when the star first appeared, it would be very brilliant. Yet there is no mention in the text of recovery of the star after conjunction with the Sun (heliacal rising) or of unusual brightness. On the other hand, near the beginning of the 8th lunar month the star was setting well after sunset in a dark sky, and under these conditions quite a faint object would be discernible.

In view of the fairly frequent incidence of errors of a month in the planetary records, it would seem most reasonable to assume such is the case here rather than draw any wild conclusions about the extreme brightness of the star based on the unquestion-ing acceptance of the duration. This implies that we are taking the period of visibility as 7 months rather than 8, although, of course, it is quite possible that the month of discovery is in error too.

Allowing for atmospheric absorption (the altitude at Nan-ching was never more than about 20 deg.), the estimated magnitude on the assumption that the object was a supernova of Type I is close to 0 at maximum. This is similar to Arcturus or Capella or Vega. Wei lies close to the direction of the galactic centre. This, coupled with the low galactic latitude of the star (the various possible remnants lie close to the galactic equator - see below), makes it probable that interstellar absorption of light would be particularly high. A distance of no more than about 5 to 6 kpc is thus suggested. "Within" (chung) the bowl of Wei, the catalogue of Clark and Caswell (1976) lists seven SNRs: G344.7-0.1, G346.6-0.2, G350.0-1.8, G350.1-0.3, G348.5+0.1, G348.7+0.3, and G349.7+0.2. The positions of these seven SNRs are shown in Fig. 6.3 along with the outline of the asterism Wei. The epoch is 1950.

On the basis of their being of low surface-brightness and small angular diameter, and therefore probably at large distance, G344.7-0.1, G346.6-0.2, G349.7+0.2, and G350.1-0.3 might initially be eliminated from the list of candidate remnants; their Σ - D distance estimates are 15.6, 13.1, 17.4, and 18.3 kpc respectively. G349.7+0.2 also has an H I absorption result confirming a distance $\geqslant$ 10 kpc. Again there remains the possibility that one or more of G344.7-0.1, G346.6-0.2, and G350.1-0.3 may be of anomalously low surface brightness, and therefore closer than the Σ - D estimates suggest - in the absence of independent distance estimates there is no way of checking for this possibility. However, even if closer than 10 kpc, on the basis of their all having small angular diameter alone none would appear to be a likely remnant for the AD 393 event.

We will now consider the three remaining candidate remnants for AD 393 in turn.

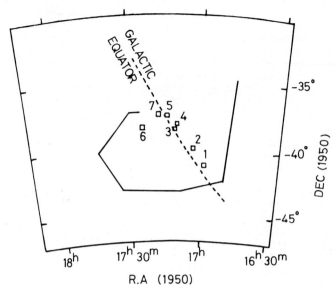

Fig. 6.3.. Catalogued SNRs within Wei.

1. G344.7-0.1	2. G346.6-0.2
3. G348.5+0.1	4. G348.7+0.3
5. G349.7+0.2	6. G350.0-1.8
7. G350.1-0.3	

The source G350.0-1.8 appears as a large angular diameter
($>$40 arc min) low surface brightness feature, which might be
only part of a much larger remnant. The map of the source from
Clark, Caswell, and Green (1975) is shown as Fig. 6.4. The
Σ - D distance is 4.3 kpc and the linear diameter $>$36 pc. The
low surface brightness and large linear diameter are suggestive
of an "old" remnant ($>$ 10,000 years), making G350.0-1.8 at best
a rather weak candidate. The expected diameter of the remnant
of AD 393 would be $\approx$15 - 20 pc; G350.0-1.8 could only have a
diameter of this order if at a distance of about 2 kpc, in contrast to the distance estimate for the AD 393 supernova of up
to 5 - 6 kpc. A distance as low as 2 kpc is very unlikely,
since a supernova this close to the Earth would be expected to
be of extreme brilliance and of long duration (as for example the
supernova of AD 185).

The two SNRs G348.5+0.1 and G348.7+0.3 are shown in Fig. 6.5;
this is the 408 MHz Molonglo map of Clark, Green, and Caswell
(1975). It has been suggested that the two remnants may be
associated because of a "bridge" linking them on high-frequency
maps (for example Milne,1969; Dickel et al,1973). The "bridge"
is also evident on the 408 MHz Molonglo map, but is not as prominent, suggesting that it has a flatter spectrum than the two
non-thermal sources; it may merely be the result of confusion
with an H II region along the line-of-sight. We will consider
the two sources to be independent SNRs, and will place no

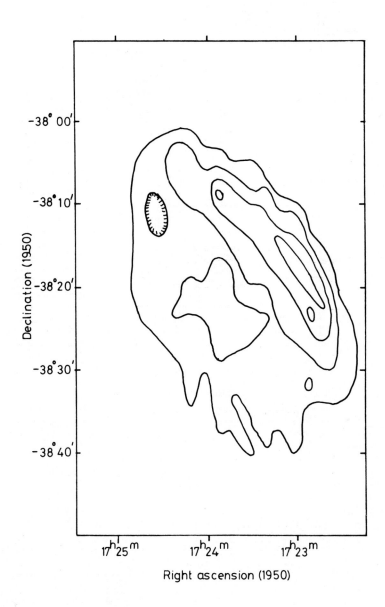

Fig. 6.4.   The SNR G350.0-1.8.

The Historical Supernovae

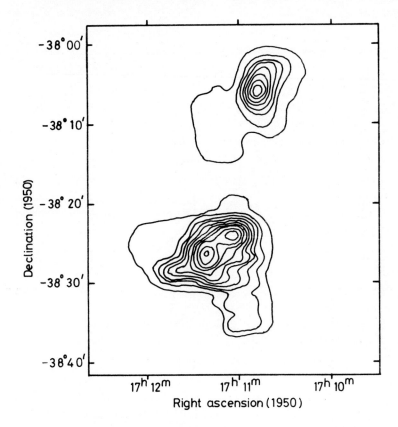

Fig. 6.5. The SNRs G348.5+0.1 and G348.7+0.3.

consequence on their positional proximity, similar surface
brightness and spectral index.

The H I absorption measurements of Caswell et al (1976) place
both G348.5+0.1 and G348.7+0.3 at a distance $\geqslant 6.7$ kpc, with
corresponding diameters of $\geqslant 16$ pc and $\geqslant 10$ pc respectively.
The lower limit is in fact slightly greater than the preferred
distance for the supernova of up to 5 - 6 kpc; however in view
of the many uncertainties involved, the discrepancy is of no
real consequence and the above data are compatible with either
source being the remnant of the supernova of AD 393. It is
unlikely that any other candidates for the remnant of this
supernova will be found in the future, since the region has now
been well surveyed.

Although there is ample vindication for the essential reliability
of the astronomical records of the Chin Dynasty, the brevity of
the descriptions of the two "long" duration guest stars of
AD 386 and 393 makes an interpretation of their exact nature and

behaviour almost impossible. The evidence for AD 386 being a
supernova remains, at best, circumstantial; the uncertain duration and brightness of the event leaves open a nova interpretation, and even if it was a supernova the vague positional descrip-
ion makes an SNR association difficult. The long period of
visibility of the AD 393 guest star means it was almost certainly
a supernova. While the positional description is essentially
an exact one, the size of the asterism Wei belies any attempt at
an unambiguous SNR identification; nevertheless, the scope for
possible equivalences is small.

Six centuries were now to pass before another supernova was seen
from Earth. It was to be the most brilliant ever witnessed.

Chapter 7

AN EXTENSIVELY OBSERVED NEW
STAR OF EXTREME BRIGHTNESS

After AD 393, there is a gap of more than 6 centuries, during
which no supernova is known to have been observed. The star
which blazed forth in AD 1006 was unquestionably a supernova,
and judging from the numerous descriptions from various parts of
the world, it was by far the brightest new star on record.

As discussed in Chapter 2, the intellectual climate in Europe
which prevailed before the Renaissance was anything but favour-
ably disposed towards the recording of a new star, and the
Arabs seem to have been scarcely more concerned with such matters.
However, because of its extreme brilliance, and despite its
southerly declination, the supernova of AD 1006 aroused a fair
amount of interest in the West, while in both China and Japan
it was very carefully observed. Indeed, not until AD 1572 do we
find another new star which is so well documented.

That we today know so much about the supernova is largely due to
the patient work of Kanda Shigeru (1935) and Goldstein, the
latter both independently (1965) and in collaboration with
Ho Peng Yoke (1965). This latter paper by Goldstein and Ho will
be referred to as GH in the rest of this chapter.

In his monumental catalogue of astronomical records of all kinds
recorded in Japanese history (often in obscure sources), Kanda
published a number of independent accounts of the star in his
section on comets and novae. Goldstein brought to light several
Arabic and European records of the new star and corrected errors
made by previous investigators. Ho Peng Yoke in GH uncovered a
number of Chinese observations and translated the Japanese
descriptions quoted by Kanda.

We have consulted all of the original Chinese and European
records of the star cited by Goldstein and GH and the more
important Japanese records, and retranslated all the important
sections. Some of the Japanese records are only of academic
interest since they tell us almost nothing about the star. In
such cases we have taken the translations direct from GH.
Neither of us knows any Arabic. Copies of the Arabic texts which
Goldstein used are provided at the end of his paper. We are
grateful to Dr. M. Y. Tamar-Agha of the University of Newcastle
upon Tyne for translating these for us. Only where there is
major discord between our own translations and those of GH or
Goldstein shall we make comment. It should be mentioned here
that in rendering Japanese and Korean accounts in this and
succeeding chapters we have preferred to give the Chinese names
for asterisms and the days of the sexagesimal cycle. We feel
that this should help to avoid possible confusion, and in any

case the Chinese characters are, of course, used in the original
texts.

The three supernova candidates discussed in previous chapters
were seen only in China. In considering the new star of AD 1006,
it thus seems appropriate to begin with the Chinese records.
These will be followed in turn by reports from Japan, Korea, the
Arab territories and Europe.

1.   China
(a)   Sung-shih astronomical treatise (chapter 56).   "3rd year of
the Ching-tê reign period, 4th month, (day) wu-yin. A chou-po
("Earl of Chou") star was seen. It appeared to the south of Ti
and 1 deg. west of Ch'i-kuan. Its form was like the half Moon,
with pointed rays shining so brightly (huang-huang) that one
could see things clearly (chien). It passed through (li) the
east of Ku-lou, and in the 8th month, following the wheel of
heaven (t'ien-lu) it entered the horizon (cho). In the 11th
month it was again seen at Ti. From this time onwards it
regularly appeared at the hour of ch'en (7 - 9 a.m.) during the
11th month at the east, and during the 8th month at the south-
west it entered the horizon".

The observations in the astronomical treatise of the Sung-shih
are classified in a variety of sections. We do not find the
record of the new star under the section entitled k'o-hsing
("guest stars"), but it is included in a separate short list of
ching-hsing ("auspicious stars"). The reason for this will
appear in due course. At certain points in the text this account
is particularly difficult to translate. We have followed GH
in rendering cho ("the turbid one" or "the turbid regions") as
"the horizon". It would be difficult to suggest a viable alternative.

In the above record, the discovery date corresponds to AD 1006
May 6. For this same year the month of heliacal setting was
equivalent to August 27 to September 24, while the star was
sighted again after conjunction with the Sun between November 24
and December 22. It is clear from the above account that the
star was seen for nearly two years at the very least, i.e. there
were at least two heliacal settings and risings. However, the
text would seem to imply a visibility for several years. This
lengthy duration immediately precludes the possibility of the
star being a comet (see Chapter 8) so that we are left with only
a nova or supernova. The use of the term li ("passed through")
in the text would seem to imply motion. However, there is ample
evidence from other sources that the star remained fixed, and
apart from the lengthy duration this is also implied by the
statement a little later in the same text that the star reappeared
at Ti. GH merely translate li as "appeared at".

(b) Sung-shih annals (chapter 7). "(3rd year of the Ching-tê
reign period) 5th month, (day) jen-yin. A chou-po star was
seen"   -   "11th month (day) jen-yin. The chou-po star was
again seen".

The above dates correspond to AD 1006 May 30 and November 26.
It seems probable that the second entry is giving us the true
date of heliacal rising of the star, since the date is at the
very beginning of the month when according to the Sung-shih
astronomical treatise the new star reappeared. Otherwise, the
annals tell us nothing. Referring to Chapter 2, we have here
an example of the gross disparity between the technically
accurate statement found in the chih and the vague notices in the
pên-chi.

(c) Sung-shih biographies. "During the 3rd year of the Ching-tê
reign period a large star appeared at the west of Ti. No-one
could determine (its significance). Some said it was a kuo-huang
or 'baleful star' (yao-hsing), which portended warfare and ill-
fortune. At that time (Chou) K'o-ming was away on a mission to
Ling-nan. On his return he urgently requested permission to
reply (to these suggestions). He said, 'I have checked the
T'ien-wên-lu and the Ching-chou-chan. The interpretation is
that the star should be called a chou-po star, which is yellow
in colour and resplendent in its light (huang-huang-jan). The
country where it is visible will prosper greatly, for it is an
auspicious star (ching-hsing). On my way back I heard that people
inside and outside the court were quite disturbed about it. I
humbly suggest that the civil and military officials be permitted
to celebrate in order to set the Empire's mind at rest. The
Emperor approved and acceded to his request. He then promoted him
to the post of Librarian and Escort of the Crown Prince .....".

This fascinating account is to be found in the biography of
Chou K'o-ming (AD 954 - 1017). More than half of the biography
is devoted to his grandfather Chou Chien, who was a great
astrologer. By his ingenuity, Chou K'o-ming averted what might
have been a calamity, and evidently deserved his promotion.
The above account is particularly interesting in that it explains
why the star was listed as a chou-po rather than as a k'o-hsing
in the astronomical treatise, and at the same time indicates that
the star had attracted widespread attention.

(d) Sung-shih biographies. "When the chou-po star appeared,
the star-clerk reported that it was an auspicious omen, and all
the officials bowed to congratulate the emperor. Chang Chih-po
gave his opinion, saying that the ruler should consolidate his
virtues in response to the signs seen in the heavens, so that
the star, which was not bound by any law, would remain visible.
He then explained the necessity of maintaining a good government.
The emperor told his officials that Chang Chih-po had the welfare
of his empire at heart".

The source of this account is the biography of Chang Chih-po.
The precise date of the events in unknown. GH point out that
another chou-po is reported in the Sung-shih annals 10 years
later and that Chang Chih-po might be referring to this.
However, the latter star is of doubtful reliability since it is
not reported elsewhere. Possibly it represents no more than a
misplaced sighting of the AD 1006 star.

(e) Yü-hu-ch'ing-hua. "During the 3rd year of the Ching-tê
reign period, a huge (t'ai) star was seen in the sky at the west
of Ti. Its bright rays were like a golden disc. No-one could
determine (its significance). Chou K'o-ming, the Chief Official
of the Spring Agency, said that according to the T'ien (-wên)-lu
and the Ching-chou-chan the star was a chou-po. It is said that
its colour is golden and its rays resplendent. Its appearance
brings peace and prosperity to the state concerned". The
Yü-hu-ch'ing-hua was written by the Buddhist monk Wêng Ying about
the year AD 1078. The above account is essentially a summary of
(c). It uses much the same phraseology and therefore cannot be
independent of it. However, the reason for this is obscure.

(f) Ch'ing-li-kuo-chao-hui-yao. "3rd year of the Ching-tê
reign period, 5th month, 1st day. The Director of the Astronomical
Bureau reported that at the first watch of the night, on the 2nd
day of the 4th month, a large star, yellow in colour, appeared
to the east of K'u-lou at the west of Ch'i-kuan. Its brightness
had gradually increased. It was found in the 3rd degree east of
Ti. Hence it belongs to the division of Chêng and the station of
Shou-hsing. The star later increased in brightness. According to
the star manuals there are four types of auspicious stars
(ching-hsing). One of these is called a chou-po; it is yellow
and resplendent and forebodes great prosperity to the state over
which it appears".

According to GH, the Ch'ing-li-kuo .... was a book presented to
the Sung emperor in AD 1044 by Chu Shu and his collaborators.
The work dealt with court matters in the period AD 960 - 1043.
The passage on the guest star exists in a preserved fragment of
the Ch'ing-li-kuo ... in a modern work, the Sung-hui-yao-kao
written by Ch'en-yüan (1935). The above description gives two
details not found in any other work - the gradual increase in
brightness of the star and the right-ascension (the third
degree east of Ti). The date of the Astronomer-Royal's report
corresponds to AD 1006 May 30, and the date of discovery of the
star in China to May 1.

GH point out that the interpretation of a chou-po star was very
subjective. In two separate places in the astronomical treatise
of the Sui-shu (chapter 20) the following descriptions of the
star may be found:
(i) It has a brilliant yellow colour and brings prosperity to
the state over which it appears".
(ii) It is large and is of a brilliant yellow colour. Over the
state it appears it presages military action, death and country-
wide famine so that the population must seek refuge from their
homes" (trans. GH).

It would appear that the precise interpretations depended very
much on circumstances and on the sagacity of the individual.
It is doubtful whether any significance should be placed on the
colorimetry, since yellow was the imperial colour of the Sung
Dynasty. (Each dynasty adopted one of the five colours blue,
red, yellow, white, black in rotation as the imperial colour).

2. Japan
(g) Meigetsuki (volume 3). "3rd year of the Kankō reign period
of Ichijō In, 4th month, 2nd day, kuei-yu. After nightfall,
within (chung) Ch'i-kuan there was a large guest star. It
was like Mars, and it was bright and scintillating. It was
seen clearly for successive nights in the south. Some suggested
that it might be due to a structural change in Chī-chên-chiang-
chün itself".

As discussed in Chapter 3, the Meigetsuki was the personal
diary of Fujiwara Sadaie, a 13th century poet-courtier, who had
a special interest in guest stars. The date given by Fujiwara
corresponds to AD 1006 May 1, precisely the same day that the
star was discovered in China - see (f) above. Several of
Fujiwara's entries concerning new stars are very detailed,
and there is no doubt that he had access to a variety of
original sources, some of them no longer extant. However, he
never specifies his source.

(h) Ichidai yoki ("3rd year of the Kankō reign period) 3rd
month, (day) wu-tzŭ amended to kêng-wu in various commentaries .
A guest star entered Chī (-kuan). Its colour was white-blue.
An instructor in astronomy, Abe Yoshimasa, reported this".

The Ichidai yoki is a Japanese record of important court events.
Its authorship and date are unknown. There was no wu-tzŭ
day in the 3rd month, and the amended kêng-wu corresponds to
AD 1006 April 28. This is three days earlier than the date in
(f) and (g), but the last day of the 3rd month was April 29
so that this amendment may be correct.

The following translations of relatively unimportant Japanese
records are taken direct from GH, with only slight amendments
(as mentioned above, Chinese names are used for asterisms and
days of the sexagesimal cycle). Following GH, unnecessary
repetition of the year (the 3rd year of the Kankō reign period)
is avoided - unless of course a different year is given.
Although of little astronomical value, these accounts give
us a fascinating insight into the general reaction of the people
at the appearance of the star.

(i) Gyokuyo (chapter 3). "The tenth day of the fourth month in
the first year of the Kao reign-period (8 May 1169). Clear sky.
Abe Yasuchika, the Assistant Astronomer, came in the evening ...
The following was said in conversation: During the time of
(emperor) Dai Sanjō In (reign: 1011 - 1016) there was an argument
between Ariyuki and Morohira. Ariyuki said that what appeared
within (the neighbourhood of) Chī-chên-chiang-chün was a 'guest
star', (but) Morohira said that it was not a 'guest star'.
Rather, he attributed this to the complete appearance of all
the stars in (the asterism) Chī-chên-chiang-chün. Morohira
found more support in their discussion and Ariyuki was much
enraged. It was prognosticated that an important national event
would take place within the next three days, and during that
time (the emperor) Sanjō In had cause for anxiety. Hence what
Morohira said was regarded as the correct explanation .....".

The <u>Gyokuyo</u> is a diary written by Fujiwara Kanezane and covers
the period AD 1164 - 1200.

(j) <u>Gonki</u>. The twenty-fourth day in the sixth month, a <u>chia-wu</u>
day (21 July 1006) .... a report on the 'guest star' was read
(before the emperor). The twenty-fifth day, an <u>i-wei</u> day (22
July) ... reports from the various routes on the guest star
were read. The third day of the seventh month, a <u>kuei-mao</u> day
(30 July) ... from behind the curtain (which separated the
emperor from his subjects) the Second Minister gave reports
from the various routes, one on the magical mirror and one on the
'guest star' - the matter concerning the 'guest star' was
still undecided. The thirteenth day in the seventh month, a
<u>kuei-ch'ou</u> day (8 August) ... (the author) went to (the palace);
a decision was made on the '(guest) star' ".

The <u>Gonki</u> is a diary written by Fujiwarano Yukinari (AD 971 -
1027), and covers the period AD 991 - 1011.

(k) <u>Hyakurensho</u>. "During the fourth month (May 1006) a large
star appeared in the SE. On the thirteenth day of the seventh
month (9 August 1006) the officials made a decision on the
reports coming from the various routes about the 'guest star' ".

The <u>Hyakurensho</u> is a history of the imperial court from AD 967
to 1259. It was written around AD 1260, but the authorship is
unknown.

(1) <u>Hōjōji Sesseiki</u>. "The thirteenth day in the seventh month,
a <u>kuei-chou</u> day (9 August 1006) ... a decision was made regarding
the reports from the various routes on the large 'guest star'
... The nineteenth day, a <u>chi-wei</u> day (15 August) ... the
cabinet ordered that divination be made regarding the large
'star'.... The eighth day of the eighth month, a <u>wu-yin</u> day
(3 September) .... offerings were made to twenty-one shrines ....
The twenty-sixth day (21 September), a <u>ping-shên</u> day .... the
Second General came and asked for an amnesty because of the
appearance of the large star".

(m) <u>Nihonkiryaku</u>. "On the thirteenth day of the seventh month,
a <u>kuei-ch'ou</u> day (9 August 1006) the lords and ministers
decided on the reports from the various routes regarding the
'guest star'. On the nineteenth day, a <u>chi-wei</u> day (21 August)
the soothsayer made divination because of the guest star. On
the nineteenth day in the eighth month, a <u>chi-ch'ou</u> day (14
September) offerings were made to the various shrines because
of the guest star".

From the above records, it is evident that the new star was
still attracting general attention as late as mid-September -
4½ months after discovery.

3. Korea
(n) <u>Koryŏ-sa</u> (chapter 47). "During the 9th year of Mokchong a
<u>hui-hsing</u> ('broom star') was seen".

This is the only astronomical record from Korea for the year
AD 1006. It is uncertain whether this has anything to do with
the new star reported in China and Japan. Although there is no
mention of a comet being observed in these countries in AD 1006,
a comet is reported in the astronomical treatise of the Sung-shih
in the previous year. It is not until AD 1011 that regular
astronomical records commence in Korea. Before then, an error
of a year in the recorded date of an event appears quite possible.
It seems best to leave the question of this Korean sighting open.

4. The Arab Dominions
(o) Alī ibn Ridwān : Commentary on the Tetrabiblos of Ptolemy.
"I will now describe a spectacle (athar) which I saw at the
beginning of my studies. This spectacle (athar) appeared in the
zodiacal sign Scorpio, in opposition to the Sun. The Sun on
that day was 15 degrees in Taurus and the spectacle (nayzak) in
the 15th degree of Scorpio. This spectacle (nayzak) was a large
circular body, $2\frac{1}{2}$ to 3 times as large as Venus. The sky was
shining because of its light. The intensity of its light was a
little more than a quarter of that of moonlight .   It
remained where it was and it moved daily with its zodiacal sign
until the Sun was in sextile with it (i.e. 60 deg. away) in
Virgo, when it disappeared at once (duf'atan wāhidatan).

All I have mentioned is my own personal experience, and other
scholars from my time have followed it and came to a similar
conclusion. The positions of the planets at the beginning of
its appearance were like this: the Sun and Moon met in the 15th
degree of Taurus; Saturn was 12° 11' in Leo; Jupiter was 11° 21'
in Cancer; Mars was 21° 19' in Scorpio; Venus was 12° 28' in
Gemini; Mercury was 5° 11' in Taurus; and the Moon's node was
23° 28' in Sagittarius. The spectacle (nayzak) occurred in the
15th degree of Scorpio. The ascendant of the conjunction when
the spectacle (nayzak) appeared over Fustat of Egypt was 4° 2'
in Leo. Also the tenth (house which included most of) Taurus
began at 26° 27' in Aries.

Because the zodiacal sign Scorpio is a bad omen for the Islamic
religion, they bitterly fought each other in great wars and many
of their great countries were destroyed. Also many incidents
happened to the king of the two holy cities (Mecca and Medina).
Drought, increase of prices and famine occurred, and countless
thousands died by the sword as well as from famine and pestilence.
At the time when the spectacle (nayzak) appeared calamity and
destruction occurred which lasted for many years afterwards ....".

Goldstein located the above text in ms. Escurial (Casiri) -
he gives a full reference. He also points out that a Latin
translation of the passage may be found in Aegidius Tebaldinus,
Quadripartitum (Venice, AD 1493). Quadripartitum is the Latin
equivalent of the Greek Tetrabiblos. This latter manuscript
gives the longitude of Mars as 21° 9' in Scorpio, and of Mercury
as 5° 5' in Taurus. Additionally the beginning of the tenth
house is given as 27° 27' in Aries. In rendering the above text,
we have followed Goldstein in translating both athar and nayzak
as "spectacle". He was here being cautious in order to avoid

prejudicing the interpretation of the phenomenon.

We shall discuss later the date of the phenomenon as determined
from the astronomical data, but Goldstein points out that a date
of occurrence in AD 1006 would fit in well with the youth of
Alī ibn Ridwān (he died in AD 1061).

(p) Ibn al-Athīr: (A.H. 396). At the beginning of Sha'ban, a
large star similar to Venus appeared to the left of the qibla
of Iraq. Its rays on the Earth were like the rays of the Moon
and it stayed until the middle of Dhū al-Qa'da and disappeared".

Ibn al-Athīr lived during the 13th century of our era. Goldstein
showed that the above account is undoubtedly taken from the
description by Ibn al-Jawzī (died AD 1200) in his book Kitāb
al-Muntazam. This is as follows :

(q) "Year 396. Among the incidents in that year a large star
similar to Venus in size and brightness glittered to the left of
the qibla. Its rays on the Earth were like the rays of the Moon.
This was on the night (preceding) Friday, the beginning of
Sha'ban, and it stayed until the middle of Dhū al-Qa'da and
disappeared".

The date of appearance of the star corresponds to AD 1006 May 3,
and the date of disappearance to around August 13 in the same year.

(r) Bar Hebraus: "A.H. 396: There appeared a great star
resembling Aphrodite in greatness and splendour in the zodiacal
sign Scorpio, its rays revolved and gave out light like that of
the Moon - it remained four months and disappeared".

The above translation from the Syriac is taken direct from
Goldstein, who acknowledges Budge (1932) as his source. Bar
Hebraus lived a little later than Ibn al-Athīr, and would appear
to be abstracting from him or Ibn al-Jawzī. At any rate, he
adds nothing new.

(s) Annales Regum Mauritaniae "In the year 396 (A.H.) a great
(variant: strange) tailed star appeared, bright (variant: burning).
(Rather long concentration was needed to see its movement).
It was one nayzak of the twelve, which was mentioned by the
ancients, and their scholars studied it over a long period.
They claimed that this star appears to a certain purpose which
God plans in the world and only God knows. (It appeared on the
first of Sha'ban, 96 mentioned above {i.e. 396 A.H.}). Its
first appearance was before sunset, then it faded until night
came and it appeared again. This star stayed for six months.
During this year it was stormy and windy, and there was thunder
without rain".

The Annales Regum Mauritaniae is an Arabic chronicle extending to
A.H. 726. Precisely the same date of first appearance is given as
in (p) and (q) above. The observation could have been made in
north-west Africa or Spain.

It would not be proper to leave this section on the Arabic observations of the new star of AD 1006 without making reference to the spurious supernova of AD 827. Humboldt (1851) in his list of historically observed new stars included an object under the year AD 827 observed by Arabian astronomers. However, he believed that the precise year was doubtful, stating, "It may with more certainty be assigned to the first half of the ninth century, when in the reign of Caliph Al Mamoun the two famous Arabian astronomers, Haly and Giafar Ben Mohammed Albumazar observed at Babylon a new star, whose light, according to their report, 'equalled that of the Moon in her quarters'. This natural phenomenon likewise occurred in Scorpio. The star disappeared after a period of four months".

This star found its way into the catalogue of Lundmark (1921), and thence into the more recent works of Hsi Tsê-tsung (1955) and Hsi Tsê-tsung and Po Shu-jen (1965). However, from the investigation of early sources, Goldstein was able to show conclusively that "Haly" was none other than Alī ibn Ridwān, and that the true date of the star was AD 1006. Tammann (1966) had gone so far as to link the supposed supernova of AD 827 with the X-ray source Sco X-3, but in a brief note, Goldstein (1966) corrected his error. Here is ample evidence that there is no substitute for consultation of original historical sources.

5. Europe
The following references are all from the compilation of monastic chronicles Monumenta Germaniae Historica (Pertz, 1826→).
(t) Annales Beneventani. "1006. In the 25th year of our master Pandolphus and the 19th year of our master Landolphus, his son, 'a very brilliant (clarissima) star shone forth, and there was a great drought for three months".

The annals of the monastery of St. Sophie, Benevento, extend from the mid-8th century AD to 1128. They are much concerned with very local events, and in all probability the star was seen in Benevento.

(u) Annales Sangallenses Maiores. "1006. A new star of unusual size appeared, glittering in aspect, and dazzling (verberans) the eyes, causing alarm. In a wonderful manner this was sometimes contracted, sometimes diffused, and moreover sometimes extinguished. It was seen likewise for three months in the inmost limits of the south (in intimis finibus austri), beyond all the constellations which are seen in the sky".

The reliable section (Pars Altera) of the chronicle of the Benedictine monastery of St. Gallen covers the period from AD 919 to 1044. Like the chronicle of Benevento, it is mainly concerned with local affairs, although it makes frequent reference to major European events. There seems no reason to doubt that the observations of the star were made in St. Gallen itself. Certainly the description, which is probably that of an eyewitness, is unique. One manuscript gives the date as AD 1012, but after AD 974 this differs by 6 years from the other manuscripts for almost all dates. Fortunately we can check which series of dates is

correct from observations of comets reported in AD 975 autumn
(981 in the discordant manuscript), 989 Feast of St. Laurence,
i.e. August 10, (995) and 998 February (1003). Oriental records
in the catalogue of Ho Peng Yoke (1962) give details of comets
in AD 975 August, 989 August and 998 February. No comets are
recorded in AD 981 and 995, and the comet of 1003 was seen in
December. It is clear that the new star which interests us was
seen in AD 1006 in St. Gallen. The record of the star is the only
entry for the year, emphasising the importance attached to it -
comparable to a famine in the previous year and a plague in the
following year.

(v) The Annales Laubienis and the related Annales Leodiensis
both mention a comet in the year AD 1006. The records run as
follows: "1006. There was a very great famine and a comet appeared for a long time."

At this period both chronicles are fairly general, covering
events throughout Europe. Presuming that the "comet" refers to
the new star, as seems likely from its long duration, there seems
no reason to believe that it was seen as far north as Belgium.
In a valuable note, Porter (1974) has pointed out that the
Belgian records may be derived from St. Gallen. He states that
a member of the influential St. Gallen family of Notker, a nephew
of Otto I, was bishop of Liege and also Abbot of Lobbes, where
the two chronicles were compiled.

(w) Iohannis Chronicon Venetum. "And so at the same time, a
comet, the sign of which always announces human shame, appeared in
the southern regions, which was followed by a great pestilence
throughout all the territories of Italy or Venice ....."

The Chronicon Venetum begins in the 6th century AD and continues to
AD 1008. The latter part of the work is generally concerned with
events in Venice itself, and it seems quite possible that John
witnessed the star he mentions. However, the precise year is not
given, and the last datable reference is June in AD 1004.
His comment, "And so at the same time ....." may suggest an
earlier event than AD 1006, although the positional description
"in the southern regions" is essentially correct for the supernova.
Our author may have been ignorant of the true nature of the star,
but in any case he is mainly concerned with what is portended.

Our attention was drawn to the following two sources by Ms. H. H.
Warwick of Minneapolis (both are to be found in Monumenta
Germaniae Historica):

(x) Alpertus de Diversitate Temporum, Lib. I. "Three years
after the king (Henry II) was raised to the throne of the kingdom,
a comet with a horrible appearance was seen in the southern part
of the sky, emitting flames this way and that".

Alpert of Metz (France) was a contemporary, and so his account
was probably original. Henry II ascended the throne in AD 1002
following the death of Otto III, so that the date of the "comet"
would correspond to AD 1005. Errors of a year are of frequent

occurrence in medieval European chronicles (c.f. Newton, 1972),
so here we may have another allusion to the supernova, with the
"flames" being the result of atmospheric turbulence at a low
altitude. A definite conclusion is not possible.

(y) Annales Mosomagenses. "1006. In this year was seen in the
sky a burning star like a torch which is called a comet".

The annals of Mousson (France) appear to be original at this period.
The year is correct for the supernova, so that we may presume
that the supernova is referred to. However, just how useful the
record is can be judged from the fact that in the same annals the
comets of AD 1066 (Halley's) and 1105 are described in exactly the
above words.

Both of these last two accounts are copied either verbatim or in a
simplified form in a wide variety of European chronicles.
Apart from the St. Gallen sighting, which more or less accident-
ally sets a valuable southern limit to the declination, the
European observations tell us virtually nothing about the star.

As far as the nature of the star is concerned, several records rule
out the possibility of it being a comet. We have the very long
duration (several years) in (a), while (a), (f), (o), (q), (s)
and (u) imply a fixed location over several months. A bright
comet moves at something like 10 deg. a day (see chapter 8) so
that its position changes rapidly from night to night. We shall
leave the question of whether the star was a nova or supernova
until a little later.

Discovery seems to have been almost simultaneous throughout the
world. In China it was first seen on a date corresponding to
AD 1006 May 1 - see text (f) - and in Japan possibly as early as
April 28 (h), but certainly by May 1 (g). One of the Arabic
reports (p), gives a date equivalent to May 3. Another report
(o) is somewhat cryptic, and deserves special discussion.

Alī ibn Ridwān gave the positions of the Sun, Moon and planets
when the star first became visible. The longitude of the Moon
is the most useful since the Moon moves at a mean daily rate of
some 13 deg. As Goldstein pointed out, Alī ibn Ridwān's data were
calculated rather than observed, as is clear from the fact that
the Moon was in conjunction with the Sun at the time and thus
invisible. However, this need not affect their reliability.
Conjunction of the Moon with the Sun occurred on April 30, and
we have interpolated the various longitudes (with the exception
of the node, which we have calculated direct) from the tables
of Tuckerman (1964). Tuckerman gave his figures for 16^h U.T.
(roughly 6 p.m. in Egypt) and we have used this time on April 30
as a preliminary moment for calculation. Alī ibn Ridwān's data,
converted to longitude in degrees and decimals, are shown for
comparison - see Table 7.1.

The position of Mars is obviously rather seriously in error, but
for the other planets and the Sun the agreement is excellent.
AD 1006 April 30 was unquestionably the date which Alī ibn Ridwān

	Alī ibn Ridwān	Modern (computed)
Sun	45°	44.9°
Moon	45	48.9
Mercury	35.18	35.45
Venus	72.47	72.58
Mars	231.32	227.87
Jupiter	101.35	100.70
Saturn	132.18	131.97
Ascending node	263.47	263.57

Table 7.1. Comparison of longitudes of the Sun, Moon and planets according to Alī ibn Ridwān and modern computation.

intended. However, if we make the necessary correction to the Moon's longitude in order to agree with the figure quoted by him, the local time corresponds to about 11 a.m. This might have been reasonable enough if the star was above the horizon, for it was bright enough to have been visible by day - see text (s). However, the account expressly states that the star was in direct opposition to the Sun, and thus well below the horizon at the calculated time. Both the position of the "ascendant of the conjunction" (i.e. the point of the ecliptic which was rising at the time) and that of the "beginning of the tenth house" (i.e. the momentary point of upper culmination of the ecliptic) are in agreement with a time around local noon. The most likely explanation that we can offer is that the star was not seen on the night of April 29 to 30, but was seen on the following night. For astrological purposes, Alī ibn Ridwān had to assume a precise time of first appearance of the new star, and thus chose noon (or a time close to this) on the intermediate day.

Before discussing the recorded duration, it is necessary to deduce the location of the star as accurately as possible. Several texts indicate a duration of around three months, and it is important to decide whether or not this is the result of heliacal setting. If not, it would imply a sharp decline in the light curve after this period.

Let us examine the records which are useful in determining the position of the new star. To begin with, the following tell us nothing useful about its location: (b), (d), (j), (k), (l), (m), (n), (s), (t), (v), (w), (x) and (y). In addition (c) and (e) are almost entirely concerned with the interpretation of the star as a portent, so that the rather vague allusion to its location ("at the west of Ti") is of little significance. We feel that the position of the star in relation to the qibla (i.e. the local direction of Mecca) in (p) and (q) is valueless in view of the uncertainty in the precise place of observation and the low accuracy with which the qibla was probably determined (Newton, 1972). Porter (1974) has discussed this last question in some detail,

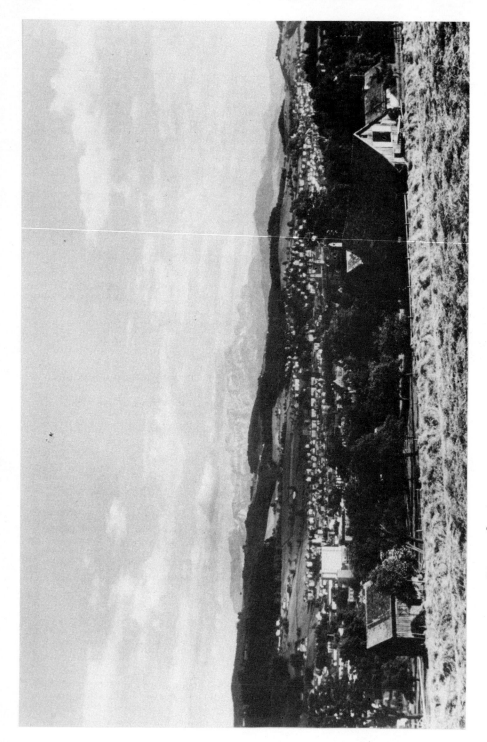

Plate 8. A view of St. Gallen and the Alpstein mountains.

and finds no alternative but to reject the observation.

The various Chinese and Japanese descriptions which are of value
all involve references to asterisms or lunar mansions. This is
a study on its own, and we shall thus commence with the Arabic
and European observations (o) and (u). Alī ibn Ridwān gave the
longitude of the star as the 15th degree of Scorpio, i.e. between
longitudes 224 and 225°, in AD 1006. Just how accurate this is
we cannot say. Comparison with the accuracy of the longitudes
of the Sun, Moon and planets which he gives cannot be made, for
these were calculated. However, as the measurement is quoted to
the nearest degree, it seems <u>reasonable</u> to assume a probable error
of a similar amount, although it would be unwise to place too
much emphasis on this.

One of the most fascinating observations is that made at St.
Gallen in Switzerland. The record, without any pretensions to
accuracy of observation, sets valuable limits to the declination
of the star, since the implication is that it was only barely
above the horizon. St. Gallen lies in a valley. The terrain
around the town is rather hilly (on the 100 m scale), but the
southern horizon is dominated by the great peaks of the Alpstein
range (maximum height 2502 m), some 20 km away. Plate 8 shows a
view of St. Gallen and the Alpstein mountains, looking towards
the south. We have used this photograph and a detailed relief
map of the area (on a scale of 1:50,000) to draw the profile of
the southern horizon as seen from St. Gallen, see Fig. 7.1.

Fig. 7.1. Southern horizon profile as seen from
the vicinity of St. Gallen.

The town itself is at an altitude of about 670 m above sea-level, while the surrounding hills reach a height of 800 or 900 m. In calculating the horizon profile, we have assumed a mean height of some 750 m in the vicinity of St. Gallen, but the precise height and location barely affect the result - we estimate that the profile is accurate to within a small fraction of a degree.

As the latitude of St. Gallen is 47.4°N, it is clear that, allowing for refraction, no star south of declination -38.5° would be visible from the neighbourhood of the town. On the other hand, if the star was as much as 5 deg. to the north of this, it could scarcely be described as so close to the horizon ("in the inmost limits of the south, beyond all the constellations which are seen in the sky"). When it first appeared, the star was in opposition to the Sun and was thus best placed for observation when on the meridian. We therefore set a somewhat arbitrary northern limit of declination as -33.5°, but it seems highly unlikely that the star was further north than this. The allusion to the severe flickering of the star would also suggest a very low altitude, where atmospheric turbulence is greatest.

Proceeding now to the Far Eastern records, Fig. 7.2 shows the area of the sky in which the new star appeared, with special reference to the oriental asterisms. Co-ordinates are for the epoch AD 1006 (precession has been applied, but proper motions are negligible). All of the stars within the figure which are brighter than magnitude +4.5 are represented. At the latitude of the Chinese and Japanese capitals (Kai-fêng and Kyōto - both close to 35° N), the meridian altitude of a star in the centre of the chart would be less than 20 deg. At such a low elevation, atmospheric dimming would be significant, and stars fainter than +4.5 would not be readily discernible. On the chart, stars in the early catalogues of Bayer and Flamsteed (which were very incomplete) are denoted by their Greek letter or number as still used today; other stars are unidentified. The western boundaries of the lunar mansions Ti and K'ang are marked for reference. Also shown are the asterisms Ch'i-kuan, Ku-lou and Chi-chên-chiang-chün mentioned in several of the texts and also other neighbouring asterisms. In drawing this chart, we have made use of the early oriental star maps discussed in Chapter 2.

We are confident that the configurations of Chi-kuan and Ku-lou in particular are correct, and there seems to be no doubt that Chi-chên-chiang-chün is the single star κ Lup.

Fig. 7.3 is more schematic. Here individual stars are not represented. The lunar mansion boundaries and the outlines of the principal asterisms are copied direct from Fig. 7.2. Also shown are the limits deduced from the European and Arabic observations (u) and (o). Let us consider the Chinese and Japanese observations in turn, referring to Fig. 7.3.

The Chinese report (a) states that the new star appeared "to the south of Ti" , probably indicating that it was within Ti lunar mansion. Later in the same text the star is described as

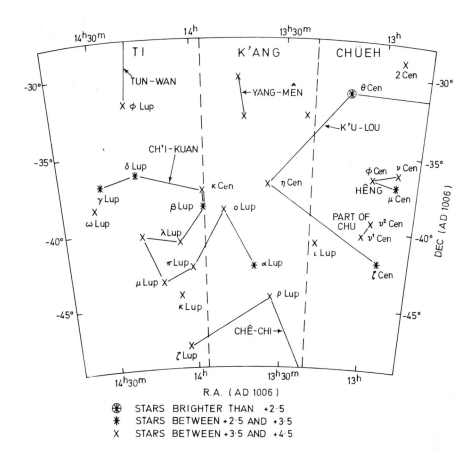

Fig. 7.2. Stars in the vicinity of the new star of AD 1006.

"at <u>Ti</u>", and this seems quite definite. However, if we interpret literally the statement that the star was "1 deg. west of <u>Ch'i-kuan</u>", this would imply a location in the western half of <u>K'ang</u>. This latter remark is not at all clear.

Fortunately (f) gives a precise right ascension - the 3rd degree of <u>Ti</u>. Coming from the Director of the Astronomical Bureau, this measurement should be reliable. Ho Peng Yoke et al (1972) quote a recent investigation by Yabuuchi (1967) of the accuracy of mid-11th century AD Chinese measurements of the positions of the determinant stars of the 28 lunar mansions. Presumably these are declinations, for the Chinese never conceived of the the idea of a single origin of R.A. (right ascensions were measured from the determinant stars themselves). The mean error found by Yabuuchi was only 0.4° with a maximum error for the sample of 1.3°. As discussed in Chapter 8, the R.A. of <u>T'ien-kuan</u> (ȝ Tau.),

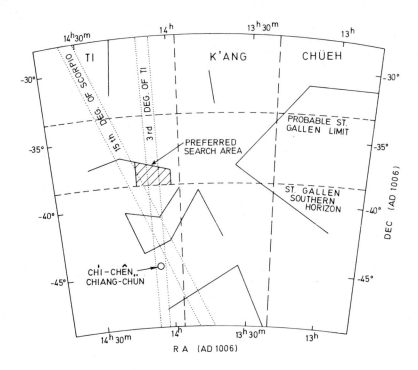

Fig. 7.3. The preferred search area for the remnant of the AD 1006 supernova.

measured at much the same time, is only about $0.9°$ in error.
It thus seems that the uncertainty in the measured R.A. of the
new star of AD 1006 should be no more than about 1 deg. Such a
location would fit in reasonably well with the description
"to the east of Ku-lou, at the west of Ch'i-kuan." In Fig. 7.3,
the zone corresponding to the 3rd degree of Ti is shown. The
agreement with the Arabic measurement (15th degree of Scorpio) is
striking.

Coming now to the Japanese records, (g) describes the star as
"within" (chung) Ch'i-kuan. We have already discussed (in Chapters
5 and 6) how precise the term chung is. Where Fujiwara obtained
his information we cannot say, but this does not alter the fact
that it is a simple matter to decide with fair precision whether
or not a new star lies within an asterism. The position of the
star thus indicated, which is confirmed in a more general way by
text (h), will be discussed below.

The possibility of the star being in the vicinity of Chi-chên-
chiang-chün (κ Lup) is ruled out by the St. Gallen sightings;
κ Lup lay fully 6 deg. to the south of the St. Gallen horizon of
the time. Fujiwara seems to have obtained his information on the
new star from two independent sources, and GH have questioned the

reliability of his reference to Chi-chên-chiang-chün. The source
of Fujiwara's information here seems to be the conversation
reported in the Gyokuyo (i). Just how accurate an account
Abe Yasuchika would have of an argument which took place some
150 years before his own time is impossible to say. GH have
pointed out that Chi-chên-chiang-chün is only a single star (the
"Cavalry General"), whereas all the stars of the asterism are
supposed to have appeared. They have suggested that Ch'i-kuan
was intended. However, the whole statement seems to be too
obscure to be of any value.

As far as the oriental records are concerned, we have so far
concentrated mainly on the R.A. of the star. However, the
several references to Ch'i-kuan, and the lack of allusion to any
asterism north of it implies a northern limit close to -35°
declination, in good agreement with the estimated St. Gallen
northern limit.

The most probable location of the star, as we define it, is
denoted by the small shaded area in Fig. 7.3. This is bounded
on the east and west by the seemingly reliable one degree uncertainty in the Chinese measurement of the ("3rd degree of Ti"),
and on the south by the firm limit set by the St. Gallen horizon.
In deciding on the northern limit, we have placed a rigorous
interpretation on the term chung in the Meigetsuki record (g).
We feel that earlier in this book we have given ample evidence
to justify this. If this particular northern limit is discounted,
the declination of the star cannot be pushed more than about 3
deg. further north since the star would have then been nearer
Tun-wan (an asterism not mentioned in any text) than Ch'i-kuan
and at the same time well above the St. Gallen horizon.

Our preferred search area could be halved if the uncertainty in
the Arabic observation (o) was limited to no more than one
degree. However, there is no supporting evidence for this, and
even at the extreme western edge of the shaded area an error of
no more than 2 deg. is implied. We therefore estimate the
co-ordinates of the star at the epoch AD 1006 as R.A. $14^h 10^m \pm$
10^m, dec. $-37.5° \pm 1.0°$ (errors are only approximate).

We are now able to discuss fully the apparent brightness of the
star at maximum and its duration of visibility. It is by no
means easy to form an estimate of the magnitude of the star at
maximum. In Japan (g), the star was compared with Mars, then in
opposition to the Sun and thus around -2. However, in Iraq
(p and q), although described as "similar to Venus in size and
brightness" (mag. about -4), "its rays on the Earth were like
the rays of the Moon". A comparison with the Moon was also made
in Egypt (o), while in China (a) and Switzerland the inference
was that it was excessively brilliant.

The key to this seemingly remarkable range of brightnesses is
perhaps supplied by a Chinese record (f) which twice mentions an
increase in brightness. Reporting on a date corresponding to
May 30 (discovery in China was on May 1), the Director of the
Astronomical Bureau stated that "the brightness had gradually

increased". In the same text we read, "The star later increased
in brightness". This suggests an increment in brightness over
several days. Presumably then, the comparisons with Mars and
Venus were made in the early stages after discovery. Certainly,
these two planets could never be compared with the Moon in brightness (mag. -12.5 at full, -10 at first and last quarters).
We know that the star of AD 1054 (see Chapter 8) reached a
peak magnitude similar to Venus, but there are no allusions to
unusual brightness although we have several records from China and
Japan. Yet the star of AD 1006, because of its extreme southerly
declination, was in a much inferior situation.

Let us consider the various accounts which mention great brightness
in order. In China, (a), the star shone so brightly (huang-huang)
that one could see things clearly (chien). The use of repetition
(e.g. huang-huang) is common in Chinese for emphasis. More
important, the character chien, which we have rendered "see clearly"
in the context of the record, generally means "to scrutinise"
or "to look closely at", implying a high level of illumination.

From Payne-Gaposchkin (1956), the starlight in the night sky
(most of which is due to faint stars) equals the light of
about 1000 first magnitude stars. However, starlight contributes
only about one-sixth of the total sky brightness; the residual
sky light comes from the zodiacal light and the "permanent
aurora". Allowing for the fact that we can only see half the
sky from any one place, and correcting for atmospheric absorption,
the integrated sky brightness is equivalent to about -7.5. To
produce the effect described, the star would clearly have to be
significantly brighter than this.

We learn from Capt. H. Haysham, of the Marine and Technical
College, South Shields, who has done much observing at sea in the
Tropics, that with the Moon high in a clear sky, one can start
to discern distant objects by its light when the age is about
5 days (elongation about 65 deg.). However, by the time the Moon
has reached first quarter, objects can be made out clearly, and
even colours begin to be distinguished. Fig. 7.4 shows the mean
magnitude of the Moon as a function of elongation from the Sun
(for elongations in excess of 30 deg.). Figures are based on data
quoted by Fessenkov (1962). From the figure, it is evident that
the 5-day old Moon reaches a magnitude close to -9, and the half
Moon nearly -10. The new star of AD 1006 never attained an
altitude of more than about 17.5⁰ at Kai-fêng, the Chinese capital
of the time, so that allowing nearly a magnitude for atmospheric
extinction, a result of magnitude -9.5 seems a conservative
estimate.

In Egypt (o), the sky was described as "shining" because of the
light of the star, and the intensity of its light was estimated
as "a little more than a quarter of that of moonlight".
It seems obvious that the full Moon was intended as the comparison
object since otherwise we would expect the phase to be stated.
Furthermore, comparison with the full Moon would be natural
since the star was in opposition to the Sun and thus closest to
the Moon when the latter was near full (less than 20 deg. away).

Fig. 7.4. Apparent magnitude of the Moon as a function of elongation.

From Muller and Stephenson (1975), changing light intensity has a subjective effect approximating the square root of the brightness. On this basis, the best estimate of the maximum brightness of the star would be about one-fifteenth of that of the full Moon, or roughly magnitude -9.5. This, somewhat surprisingly for such crude estimates, agrees with our previous result. According to Goldstein, a possible reading of the text is that the star was "as bright as the Moon a little more than one-quarter illuminated". The Moon at this phase has a magnitude close to -8.5. However, correcting for differential atmospheric extinction, since the Moon at first quarter was in a high northern declination, we return to -9.5 for the star.

The mention of the apparent angular size of the star in the same text ("$2\frac{1}{2}$ to 3 times as large as Venus") is difficult to interpret. Due to an optical effect whose cause is partly physiological, all stars appear to have an apparent diameter which is greater the brighter the star. Thus Tycho Brahe estimated the apparent diameters of stars of different magnitudes as follows: 1st mag. 120", 2nd 90", 3rd 65", 4th 45", 5th 30", 6th 20". Again, the mediaeval Arabian astronomer Al Fargani gave the apparent diameters of the planets as : Mercury 1/15 of the Sun, Venus 1/10, Mars 1/20, Jupiter 1/12, Saturn 1/18 (for both sets of data

see Dreyer, 1906). To attempt to extrapolate on these figures in
order to deduce the apparent magnitude of the new star would be
mere guesswork, especially in view of the close agreement
between the figures for Jupiter and Venus, but the large estimate
of the apparent size of the star suggests that it was very much
brighter than Venus.

From Iraq (p and q) we find a further independent comparison with
the Moon, while at St. Gallen, although the light loss on account
of atmospheric absorption at such a low altitude would be some
2 mag., the star could still be described as "dazzling" or
"beating at" the eyes. All of the various photometric indications
are necessarily very rough, but we feel that the best estimate
of the apparent magnitude of the star at maximum is within 1 mag.
of -9.5. (It is interesting to note that Goldstein deduced
-8 to -10). A star fainter than -8.5 could scarcely produce
the effects described, while if much brighter than -10.5 it
would rival the full Moon.

The absolute magnitude of a nova varies from -8.5 (fast) to
-6.5 (slow)-see Chapter 8. Thus if the star was a fast nova it
would be at a distance of less than 10 pc from us, and if a slow
nova less than 5 pc away. It would therefore be among the very
nearest stars. The probability of such an event occurring during
the last 2,000 years (the period of detailed recorded history)
is very small. In fact, a fair estimate of the frequency of
novae occurring less than 10 pc away and as close to the galactic
equator as the new star of AD 1006 (some 15 deg.) is about once
every million years.

Additionally, the possibility of the star being a fast nova seems
to be excluded by the rather careful Chinese description (f)
of its gradual increase in brightness. Fast novae tend to rise
to a maximum within a day or so and then decline rapidly
(c.f. Payne-Gaposchkin, 1957). The light curve of a slow nova
follows no definite form, but the remnant, a white dwarf in a
binary system, is typically less than 10 mag. fainter than the
maximum of the outburst. In the event of the star being a slow
nova, the remnant would be close to mag. 0. The brightest star
shown in Fig. 7.2 is θ Cen. (mag. +2.3). All the stars represented in the figure are at a distance of more than about 50 pc
with the exception of θ Cen. and 2 Cen., which are both about 20
pc away (c.f. the catalogue of Schlesinger, 1940). However, both
of these stars are isolated red giants, and in any case lie two
lunar mansions to the west of Ti and thus can be discounted.

In the catalogue of trignometric stellar parallaxes compiled by
Jenkins (1952), there is not a single star listed closer than
50 pc down to mag. +9, within 5 deg. of the centre of our preferred search area. This catalogue seems fairly complete down
to mag. +9, and contains many stars fainter than this. A
star as close as (or closer than) 10 pc would be expected to have
a high proper motion (of the order of 1" annually), so unless
extremely faint, the likelihood of it being missed seems very small.

The evidence is thus in favour of the star being a supernova.
With an apparent magnitude close to -9.5, a maximum distance of
roughly 1 kpc is indicated. (At such a high galactic latitude
interstellar absorption would be negligible).

Our final preliminary question concerns the recorded period of
visibility of the star. Several texts mention a visibility of
roughly 3 months. Alī ibn Ridwān is quite specific when he
states that the star disappeared "at once" when the Sun came
within 60 deg. of longitude of it. However, by this time,
from the latitude of Fustat (30ºN) the star would be only above
the horizon during the hours of daylight. After such a period
both a Type I and a Type II supernova would have faded by some 4
magnitudes so allowing for atmospheric absorption the star would
appear no brighter than Venus. The visibility of this planet
in daylight is very subjective.

Much the same situation would hold in Iraq and China, so that
there is a simple enough explanation for the disappearance of
the star. However, it does not seem possible to decide on the
type of supernova. Lacking extensive data on Type II light
curves (see Chapter 8), the lengthy total duration of the star
recorded in China (a) cannot be regarded as positive evidence of
Type I behaviour.

At such high galactic latitude ($\sim 15^{\circ}$), interstellar absorption
would be low. Thus a peak apparent magnitude of -9.5 would
suggest a distance to the supernova of $\sim$1 kpc. A distance as
great as 1.3 kpc as suggested by Minkowski (1966) would only be
possible assuming zero interstellar absorption, Type I properties,
and peak apparent magnitude $\sim$-8.5. We will assume a distance of
1 kpc with uncertainty of about 0.3 kpc.

A radio search of the region of interest was reported by Milne
(1971), and his 635 MHz map is shown as Fig. 7.5. The source
PKS 1459-41 (G327.6 + 14.5), diameter 34 arc min, is one of two
SNRs within the region. The other is the Lupus Loop (G330.0 + 15.0)
a large arc ($\sim 3^{\circ}$ radius) of very low surface brightness. The
radio properties are suggestive of an SNR of extreme age
($\sim 10^{5}$ years old), distant about 0.5 kpc and with diameter $\sim$50 pc.
The positions of the two SNRs for epoch AD 1006 are shown in
Fig. 7.6. The question arises as to whether the Lupus Loop
might be a possible remnant of the AD 1006 event, despite its
radio properties. If it were at the estimated distance of about
1 kpc for the supernova, then the mean expansion rate of the
remnant would need to be >150,000 km s-1 - an absurdly high
value! To reduce this to an acceptable mean expansion rate for
a 1000 year old supernova remnant would require it to be at a
distance of no more than 100 pc. Such a nearby outburst would
occur on average once per 200,000 years, with an expected
absolute magnitude of about -14 or brighter! An object many
times brighter than the full Moon would surely have attracted
universal attention, and such extreme brightness is certainly not
in accord with the historical records of AD 1006. On these
grounds, quite apart from the fact that the remnant lies well
outside our preferred search area, the Lupus Loop must be entirely

Fig. 7.5. A 635 MHz radio map of the Lupus
region (from Milne, 1971).

discounted as a possible remnant of the AD 1006 event.

As noted in Chapter 4 the SNR PKS 1459-41 has the highest galactic latitude of any 'young' remnant, and is only one of four known
SNRs lying beyond 10° of the plane. It lies on the periphery of
our preferred search area. The chance positional coincidence of
a historical new star and a young SNR at such galactic latitude
would be extremely remote (less than 1 in 1,000); the available
evidence must favour a definite supernova classification for the
AD 1006 new star, with PKS 1459-41 as its remnant. This association fits the St. Gallen observations particularly well, and
the apparent path of a star at the location of PKS 1459-41 for
AD 1006 is shown in Fig. 7.7.

The radio source PKS 1459-41 shows the broken-ring structure
characteristic of an SNR (see Fig. 7.8); in addition, a clearly
identified non-thermal spectrum (α =-0.57) and significant
polarization from regions of peak radio brightness confirm a
synchrotron origin for the emission. The direction of polarization suggests that the magnetic field is radial, and for a
'young' SNR such a field distribution may occur as a result of

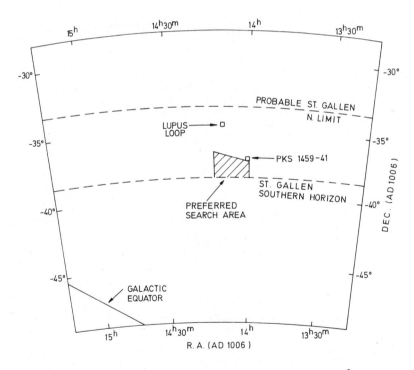

Fig. 7.6 SNRs in the vicinity of SN 1006.

Fig. 7.7. Apparent path of a star at the
location of the SNR PKS 1459-41 in relation
to the St. Gallen horizon in AD 1006.

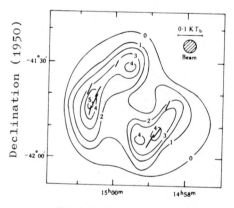

Right ascension (1950)

Fig. 7.8. A 5000 MHz map of the remnant of the
AD 1006 supernova (from Milne, 1971).

radial stretching of inherent field lines in the expansion of the
ejecta. As will be discussed in Chapter 12, the source is of
unusually low surface-brightness for its age; however there are
other such sub-luminous SNRs, and there are no serious objections
to an association of PKS 1459-41 with the AD 1006 supernova.

The optical detection of the remnant has only recently been
reported by van den Bergh (1976, IAU Circular No. 2952). Using
the 400 cm reflector at Cerro Tololo Interamerican Observatory,
delicate wisps of filamentary nebulosity were detected on the
north-west periphery of the radio source. Individual filaments
are of about 9 arc min in length and from 1 to 8 arc sec in
thickness. Van den Bergh suggests that the morphology of the
filaments is very reminiscent of that of the optical remnant of
the Type I supernova of AD 1572 (see Chapter 10); however as yet
it is not possible to confirm this classification on other
grounds.

Winkler and Laird (1976) have reported the positive detection of
the SNR with the 1 - 10 keV X-ray instrument on the satellite
OSO-7; however this detection in X-rays has yet to be confirmed.
Nevertheless, on the basis of a thermal origin for the X-ray
emission the OSO-7 data suggests a distance of 1.2 kpc to the
source, consistent with the distance estimated for the AD 1006
supernova from the historical accounts of its brightness.

It is unfortunate that the remnant of the most spectacular stellar
outburst ever recorded by man should have been so poorly studied
in modern times. Radio observations have been rather limited,
and the optical and X-ray emission only recently detected.
Theoretical consideration of the evolution of the source seems
almost non-existent. We would completely discount criticism of
the classification of the AD 1006 outburst as a supernova, and
believe that the association with the SNR PKS 1459-41 must now
be taken as certain. It is to be hoped that general acceptance
of this association will mean that in future the source will not
suffer the observational and theoretical neglect of the past.

Chapter 8

THE BIRTH OF THE CRAB NEBULA

The new star which appeared in AD 1054 was not nearly so bright as its predecessor almost half a century earlier. However, once again we find an impressive list of observations from the Orient.

More than 30 years ago, a detailed survey by Duyvendak (1942) brought to light 5 separate Chinese and Japanese observations of the star. Scarcely any further records have since been discovered, although in recent years the known observations have been subject to much closer scrutiny.

The Crab Nebula, the remnant of the star, has become one of the most extensively studied objects in astrophysics, but it is perhaps useful to get the original outburst in true perspective. Although well placed for observation in the Northern Hemisphere, there is not a single known record of the star from Europe or the Arab Lands. A Chinese record (see below) implies that the star remained visible for no more than 23 days in daylight, suggesting that it was never much brighter than Venus. It was thus considerably fainter than the new stars of AD 185 and 1006, and was probably of similar brilliance to the supernova of AD 1572 (see chapter 10). Nevertheless, the impact that a number of crude observations made nearly a thousand years ago has had on astrophysical research today seems scarcely believable.

Let us look at the oriental records of the new star. These are all from China and Japan; no astronomical observations of any kind are reported from Korea (in the Koryŏ-sa) in AD 1054 (the 8th year of King Munjong). The reason for this is obscure since the previous year is well covered. Possibly the original observations were lost before the Koryŏ-sa was compiled (there are numerous similar gaps at other periods).

Translations of the Chinese and Japanese records are as follows:

(a) Sung-shih (Astronomical Treatise, chapter 56). "1st year of the Chih-ho reign period, 5th month, (day) chi-ch'ou. (A guest star) appeared approximately several inches (k'o-shu-ts'un) to the south-east of T'ien-kuan. After a year and more it gradually vanished (mo)".

The date of first appearance corresponds to AD 1054 July 4. This report is to be found in the section entitled "Guest stars" in the astronomical treatise, hence the nature of the star is omitted in the text itself. This same account is to be found verbatim in the section on guest stars in the Wên-hsien-t'ung-k'ao, an encyclopedia compiled by Ma Tuan-lin around AD 1280. As this was more than half a century before the Sung-shih was composed,

it is apparent that Ma Tuan-lin had access to the same source as
the writers of the official history.

(b) Sung-shih (Annals, chapter 12).   "(1st year of the Chia-wu
reign period, 3rd month) (day) hsin-wei. The Director of the
Astronomical Bureau reported that since the 5th month of the 1st
year of the Chih-ho reign period a guest star had appeared in
the morning at the east, guarding (shou) T'ien-kuan, and now it
has vanished (mo)".

This report confirms the month of first sighting given in (a)
above and furthermore gives the exact date when the star was no
longer visible. This latter date corresponds to AD 1056 April
17, making the total duration more than 21 months. It is interesting to note that this is the only mention of the star in the
Annals; there is no reference to it in the discovery year.
We might remark that we have here by far the most valuable account
of a new star in the Pên-chi (i.e. Imperial Annals) of any
dynastic history.

(c) Hsü-tzǔ-chih-t'ung-chien-ch'ang-pien (chapter 176).   "(1st
year of the Chih-ho reign period, 5th month), (day) i-ch'ou.
A guest star appeared approximately several inches to the south-
east of T'ien-kuan".

The Hsü-tzǔ-chih ..... is a Sung chronicle written by Li Tao, who
died in AD 1184 It thus precedes the Sung-shih by more than 150
years, but even so a similar lapse of time had occurred between
the observation of the new star and Li Tao's recording of it.
Except that it does not mention the disappearance of the star,
the text is virtually identical with that in the Astronomical
Treatise of the Sung-shih (a) above. Indeed there are only two
slight differences. The Hsü-tzǔ-chih contains an additional character chih, an optional connecting link which does not alter the
sense in any way. However, this work gives the cyclical day as
i-ch'ou (the 2nd day) of the Sung-shih. Reference to chronological
tables (e.g. those of Hsüeh Chung-san and Ou-yang I, 1956) shows
that there was no i-ch'ou day in the 5th month of the discovery
year, but there was a chi-ch'ou day. However, the written
characters chi and i are so similar that they are readily confused.
Ho Peng Yoke (1966) cites many such instances occurring in the
Chin-shu. Towards the end of the 5th month, the Hsü-tzǔ-chih
has successive entries on the 18th, 22nd and 23rd cyclical days,
followed immediately by the guest star account under the day
i-ch'ou. It seems clear that Li Tao originally had the guest
star listed on the day chi-ch'ou (the 26th day), but this was
subsequently corrupted to i-ch'ou.

A late summary of the work, the Hsü-tzǔ-chih-t'ung-pien gives the
correct cyclical day chi-ch'ou, but without comment. So much for
this textual question.

(d) Sung-hui-yao (chapter 52). "1st year of the Chih-ho reign
period, 7th month, 22nd day ..... Yang Wei-tê said, 'I humbly
observe that a guest star has appeared; above the star in question there is a faint glow, yellow in colour. If one carefully

examines the prognostications concerning the emperor, the interpretation is as follows: The fact that the guest star does not
trespass against Pi and its brightness is full means that there is
a person of great worth. I beg that this be handed over to the
Bureau of Historiography'. All the officials presented their
congratulations and the Emperor ordered that it be sent to the
Bureau of Historiography.

During the 3rd month of the 1st year of the Chia-yu reign period
the Director of the Astronomical Bureau said, 'The guest star has
vanished (mo), which is an omen of the departure of the guest'.
Earlier, during the 5th month in the 1st year of the Chih-ho
reign period (the guest star) appeared in the morning in the east
guarding (shou) T'ien-kuan. It was visible in the daytime, like
Venus. It had pointed rays on all sides (lit. 'in the four
directions') and its colour was reddish-white. Altogether it
was visible for 23 days".

The Sung-hui-yao ("Essentials of Sung History") was compiled by
Chang Tê-hsiang at some time during the Sung Dynasty. The account
of the guest star is clearly in two parts, one of which is mainly
astrological and the other which gives valuable astronomical
details. The first section is very difficult to translate.
It seems possible that Yang Wei-tê was using the appearance of the
guest star to further his own political motives. This is particularly apparent in the oblique reference to Pi, an asterism more
than 15 degrees from T'ien-kuan (see below). His allusion to
the colour of the star may be unreliable since as mentioned in
Chapter 7, yellow was the imperial colour of the Sung Dynasty.
Yang Wei-tê made his speech nearly two months after the appearance
of the new star; the date corresponds to AD 1054 August 27.

In the second part of the account, the months of first appearance
and final disappearance of the star recorded in the Sung-shih -
(a) and (b) above - are confirmed. As the total period of
visibility was nearly two years, and the star was not long past
conjunction with the Sun when it was first seen (see below),
the duration of 23 days must refer to the visibility of the star
in daylight. Sightings of Venus in daylight are often recorded in
Far Eastern history and it is fairly usual to state the exact
number of days that the planet was thus visible.

(e) K'i-tan-kuo-chih (chapter 8). "(23rd year of the Chung-hsi
reign period), 8th month, the king died ...... Previously there
had been an eclipse of the Sun at midday, and a guest star appeared
at Mao. The Deputy Officer in the Bureau of Historiography,
Liu I-shou, said, 'This is an omen that (King) Hsin-tsung will die'.
The prediction indeed came true".

The K'i-tan-kuo-chih ("Memoirs of the Liao Kingdom"), was written
by Yeh Lung-li around AD 1350. Liao was a semi-nomadic kingdom,
existing in the extreme north of China from AD 937 to 1125.
Duyvendak (1942) pointed out that the king died in the year after
the stated year, on a date corresponding to AD 1055 August 28,
but he dated the solar eclipse as AD 1054 May 10. This was total
in Central China (cf. Oppolzer, 1887) and would be fairly large in

the north (there was no other large eclipse until AD 1061). We
can obviously no more than guess at the date of the guest star,
but judging from the eclipse date it would seem to fit in reasonably well with that in the Sung records. However, Mao is some
20 degrees from T'ien-kuan (for further discussion see below).

Ho Peng Yoke et al (1972) point out that the astronomical details
of the above record are copied verbatim in a late work, the
Sung-shih-hsing-pien, written about AD 1550, but presumably the
compiler had access to the Liao account.   N.B. as the work by
Ho Peng Yoke et al just cited will be alluded to several times in
this chapter, we shall subsequently abbreviate it to HPP .

(f) Meigetsuki (volume 12).   "2nd year of the Tenki reign period,
4th month, after the middle decade. At the hour ch'ou (1 - 3 a.m.)
a guest star appeared in the degrees of Tsui and Shen. It was
seen in the east and flared up (po) at T'ien-kuan. It was as
large as Jupiter".

We have already mentioned the Meigetsuki, the diary of the 13th
century AD Japanese courtier Fujiwara Sadaie, in chapter 7.
The account of the star is to be found under the entry for the
8th day of the 11th month of the Kwanki reign period (AD 1230
December 13). The date of first sighting of the star given in
the text corresponds to the days immediately following AD 1054
May 29, but this clashes with the Chinese date. This point will be
considered in detail below.

In the above translation we have rendered po as "flared up".
It is evident that a verb is intended here. HPP supposed that
po implied the compound po-hsing (i.e. a tail-less comet, see
Chapter 3), and thus they inferred two separate stars. However,
this is highly improbable - we would have expected the full
compound, and in any case use of the verb is quite common.
Additionally, T'ien-kuan lies "In the degrees of Tsui and Shen" -
see below - so that it is clear that the text refers to one and
the same star. This was realised by Duyvendak (1942).

(g) Ichidai yoki (volume 1). "2nd year, chia-wu (of the Tenki
reign period), 4th month. A great star appeared in the degrees
of Tsui and Shen. It was seen in the east and flared up at
T'ien-kuan. It was as large as Jupiter".

The Ichidai yoki is a Japanese chronicle of unknown date and authorship. In the original account there are two examples of textual
corruption involving very similar characters. These errors were
corrected by Kanda (1935). The characters tu ("degrees") and
kuan (in T'ien-kuan) are replaced by characters which closely
resemble them, but the mistakes are obvious. Otherwise, the
Ichidai yoki record is identical with that in the Meigetsuki -
(f) above. As the Meigetsuki gives more information (regarding
date and time), and is free from error, it is in all probability
the earlier source. In any event, the two records are by no
means independent, as for (a) and (c) above.

(h) Dainihonshi (chapter 359). "2nd year of the Tenki reign
period, 4th month, A guest star was seen". This very brief
account in such a late Japanese history may well be derived from
either (f) or (g).

Summarising the historical sources, we have 5 independent accounts,
(a), (b), (d), (e) and (f). As already pointed out, the Chinese
sources (a) and (c) share a common origin, as is true of the
Japanese sources (f) and (g). In each case the former gives
more information so that we shall discard the latter in subsequent
discussion. Our object is now to collate the various historical
evidence concerning the new star.

Until recently, it was widely accepted that the new star of AD 1054
was a supernova and that the Crab Nebula was its remnant.
However, HPP criticised both conclusions. Accordingly, we have
felt it desirable to subject the historical material relating to
the star to particularly detailed analysis. Leaving a discussion
of the exact nature of the star for the present, let us begin
by considering the asterisms mentioned. Of the independent sources
selected above, (a), (b), (d) and (f) all mention T'ien-kuan.
It is thus important to establish its identity. Mayall and Oort
(1942) in their discussion of the Crab Nebula, believed that it
was not possible to locate the asterism "much more precisely than
'near ζ Tauri'". However, HPP made a detailed investigation
of early Chinese historical sources and proved conclusively that
T'ien-kuan is ζ Tau itself, i.e. an isolated star. Judging from
the name ("Celestial Gate") two stars would be expected. Of
the many works which HPP cite, we have selected two which are most
specific, the familiar Chin-shu and the Ling-t'ai-pi-yüan ("The
Secret Garden of the Observatory"). This latter work, originally
written by Li Chi-tsai about AD 580, was revised around AD 1050
by Wang An-li, a member of the Sung Astronomical Bureau to
incorporate observations made in his own time.

According to the Chin-shu, "T'ien-kuan is a single star south of
Wu-chê. It is also called T'ien-mên (which also means "Celestial
Gate"). It lies where the Sun and Moon move".

The Ling-t'ai-pi-yüan is more specific. We read, "T'ien-kuan is
a single star north of Tsui. T'ien-kuan is distant 71' degrees'
(tu) from the Pole and is within the 1st degree of Tsui". HPP
point out that Ma Tuan-lin in his encyclopedia, the Wên-hsien-
t'ung-k'ao, quotes a contemporary Sung source, no longer extant,
which gives the polar distance as $71\frac{1}{2}$ degrees - otherwise the
description of position is identical with that in the Ling-t'ai-
pi-yüan. Somewhere a copyist's error has occurred. We should
mention here that 1 tu is equivalent to 0.9856 degrees; there are
365.25 tu to a full circle.

The most useful detail of the Chin-shu description is that
T'ien-kuan lies on (or close to) the ecliptic. Tsui, mentioned in
the Ling-t'ai-pi-yüan, is one of the lunar mansions and its
determinant star is λ Ori (see Chapter 2). HPP point out that
between AD 1049 and 1053 the co-ordinates of many stars were
determined or checked by the Sung astronomers using an armillary

sphere;   this is possibly the period when the position of T'ien-kuan,
as recorded in the Ling-t'ai-pi-yüan, was measured.   It would then
appear that around the epoch AD 1050 the RA of T'ien-kuan was perhaps 0.5°±0.5° in excess of that of λ Ori, and the declination
either +21.0°±0.5° or +20.5°±0.5°.   Applying precession, HPP
have reduced the co-ordinates to the epoch AD 1950.   The RA is
then about 5^h 38^m and the dec.+21.7° or 21.2°, with an uncertainty
of about 0.5 degrees in each co-ordinate.   This position is in
excellent accord with that of 3 Tau itself (viz. $5^h34.7^m$, +21.1°).
As this star is fairly bright (magnitude +3.0), and there are no
other stars significantly brighter than magnitude +5 within as much
as 8 degrees of it, we can regard the evidence for the identity
of 3 Tau with T'ien-kuan as conclusive.

The various oriental star maps discussed in Chapter 2 in every
case show T'ien-kuan as a single star, and from the more accurate
maps, especially the Jesuit and Su Sung charts, it is clear that
the location is the same as that of 3 Tau.   Thus we have additional
independent evidence (if any were needed).

As we have seen, the Chinese records (a), (b) and (c) mention
T'ien-kuan only - in (d) there is an oblique reference to Pi
(one of the constituents of which is α Tau).   On the other hand,
the Japanese record (f) additionally states that the star appeared
in "the degrees of Tsui and Shen".   These are two adjacent lunar
mansions - the 20th and 21st (see Chapter 2).   Fig. 8.1 shows the
position of T'ien-kuan in relation to Tsui and Shen, and the
neighbouring lunar mansion Pi.   The boundaries of the zones of
RA defined by Tsui are marked.   These correspond to the positions
of the determinant stars of Tsui and Shen.   The position of the
Crab Nebula is also shown.   To illustrate the relative isolation
of 3 Tau, we have shown all stars brighter than magnitude +4.5.
From the diagram, it is clear that the Japanese description is
quite accurate within its own limitations.

The remaining account (e) from the Liao kingdom mentions only the
asterism Mao, i.e. the Pleiades.   This is more than 20 degrees
from 3 Tau, and it is therefore questionable whether the same
star is referred to.   However, the Chinese and Japanese records
do not mention any other new star near this time.   Further, the
date of the solar eclipse mentioned in (e) ties in well with the
date of appearance of the star near T'ien-kuan.   As the record is
mainly concerned with the death of King Hsin-tsung, rather than
astronomical technicalities, the writer may have given simplified
information.   The Pleiades are one of the few star groups with
which everyone is familiar.   If we had had only a single reference
to T'ien-kuan, then the discord with the Liao account might have
been serious.   However, we have ample independent evidence that
the new star appeared near T'ien-kuan.   We shall subsequently
confine ourselves to the three Chinese records (a), (b) and (d),
and the Japanese record (f).

Having considered the asterisms mentioned in the various texts
and shown that there is general agreement on a single position
close to 3 Tau, we may now discuss the nature of the star.   Let
us begin by considering the exact period of visibility.

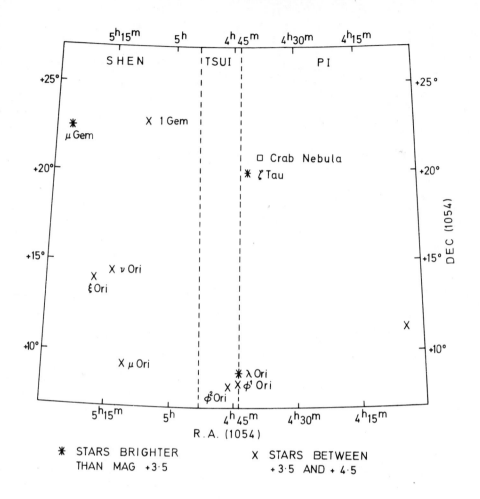

Fig. 8.1. Stars in the vicinity of ζ Tau.

Taking the Chinese records first, (a) gives an exact date of
first sighting - the day chi-ch'ou of the 5th month of the Chih-ho
reign period, corresponding to AD 1054 July 4. Accounts (b)
and (d) confirm the lunar month as the 5th month. On the other
hand (a) states only that the star was visible "for a year and more"
(which could mean anything up to two years), while (b) gives an
exact date of last visibility - presumably the day before hsin-wei
of the 3rd month of the Chia-wu reign period, i.e. AD 1056 April 16.
We have found in earlier chapters that when Chinese dates are given
to the day they are generally correct, and there seems to be no
reason for questioning the first and last sighting of the star.
For an object which is visible for more than a year, conjunction
with the Sun must occur, but we find no mention of heliacal setting
or rising. However, there are no means of knowing how much

information relating to the star was discarded by the official
historians when the Sung-shih was written up.

When we turn to the Japanese record (f), we find only a mention of
the first sighting, but the date given is much earlier than that
in China. The last day of the middle 10-day period in the 4th
month of the Tenki reign period corresponds to AD 1054 May 29.
According to the text, the star became visible in the days immediately following. Yet it seems inconceivable that the star could
have been sighted in Japan a full month before its discovery in
China. Additionally, as was pointed out by Duyvendak (1942),
₃ Tau would be in conjunction with the Sun around the time of the
recorded Japanese sighting (the precise date was May 27).
Duyvendak preferred to assume an error of a month in the Japanese
record, reading "5th" for "4th", in which case the date of first
sighting in Japan would be June 30 onwards. This seems the most
reasonable explanation (we have frequently noted errors of a month
in earlier chapters). However, as the Chinese dates of discovery
and disappearance are precisely given we shall adopt these as
definitive.

As far as the nature of the star is concerned, it is significant
that both the Chinese and Japanese records avoid the use of the
normal terms for a comet. The object was described as a guest
star (k'o-hsing) in China and Japan, and this could refer to either
a nova (or supernova) or a comet without a perceptible tail -
see Chapter 3. Further, the various accounts remain silent on any
motion of the star during the very long period of visibility
which would be unusual for a comet (nearly two years). That
the star was still in the vicinity of ₃ Tau nearly two months after
its first appearance is shown by the account (d) in which the
Astronomer Royal discussed the astrological implications of the
observation that the star did not "invade" the bright asterism
Pi. This star group is less than 15 degrees from ₃ Tau.

Both (b) and (d) describe the guest star as "guarding" (shou)
T'ien-kuan. Ho Peng Yoke (1966) lists several early definitions
of this term, e.g. "to attach and stay by the side of (a constellation or celestial body)", or "to stay (in a particular place)
without leaving". Both of these definitions indicate a fixed
position. On the other hand HPP, who express doubts regarding
the nature of the star, give some importance to an alternative
interpretation of shou - "moving backwards and forwards without
going away". Such an apparent motion could result from the
revolution of the Earth, which completed nearly two full orbits
while the star was visible, but only if the latter was stationary.
We may conclude that the guest star either showed no apparent
motion or at most a slight parallactic displacement.

If the new star of AD 1054 was indeed a comet, the extended period
of visibility, absence of a tail and lack of any significant
motion cannot be reconciled with the observation (d) that it was
seen for 23 days in daylight. Daylight comets are by no means
infrequent (cf. Chambers, 1909), but they are characterised by
a lengthy tail (several tens of degrees), rapid apparent motion
at brightest (some 10 degrees daily), short duration in daylight

(usually only two or three days at most) and short period of
visibility to the unaided eye (a few weeks or months). Ho Peng
Yoke (1962) gives detailed translations of oriental records of
several hundred comets observed during the last 2500 years.
A few of these, described as hui-hsing ("broom stars") or po-hsing
("rayed stars") were seen for 5 or 6 months, but this is the limit
- the maximum recorded duration is no more than 190 days (in
AD 254).

Various independent lines of evidence thus concur to disprove a
cometary interpretation of the star. Accordingly, we must infer
that the object was either a nova or supernova. The "pointed
rays" described in (d) as emanating from the star may then be
explained as the result of distortion within the eye. This
phenomenon is commonly observed with Venus, especially near
maximum brightness.

Having established the stellar nature of the new star, we must
consider its position in relation to 3 Tau in order to obtain
accurate co-ordinates. We find its location described in (a) as
"approximately several inches (ts'un) to the south-east of T'ien-
kuan" and in (f) as "at T'ien-kuan" (where "at" must be understood as there is no character in the text). Obviously the star
must have appeared in close proximity to   3 Tau.The use of a linear
unit (ts'un) requires clarification. For some reason or other, in
Far Eastern astronomy angular separations are regularly quoted in
linear units - either ts'un, ch'ih (equal to 10 ts'un) or ch'ang
(equal to 10 ch'ih). The first of these is rather smaller than the
British inch. As Kiang (1971) so well remarks, "It may be observed
that although in tu (i.e. degrees of 365.25 to a circle) we have
a perfectly good angular unit, it is never used to express the
distance between heavenly bodies, or the lengths of comet's
tails, meteor trails or zodiacal light. Instead these linear
units are invariably used in such cases".

On the assumption that in the record of the guest star designated
as (a) above, the Chinese astronomers were referring to the scale
of their armillary sphere, HPP attempted to derive the angular
equivalent of the ts'un in the text. The armillary sphere constructed by the Chinese astronomer Han Hsien-fu around AD 1000
had a circumference of 18 ch'ih (180 ts'un). Judging from the list
of the dimensions of such spheres throughout Chinese history given
by Needham (1959), this figure seems to have been fairly typical.
Obviously, on such an instrument 1 ts'un would correspond to
about 2 degrees.

However, this hypothetical result cannot be correct. Kiang
(1971) made a valuable study of 7th century AD Chinese measurements
of the position of the ecliptic in relation to the 28 lunar mansions. He derived the result 1 ch'ih = $1.50° + 0.24°$ (s.e) .
Thus 1 ts'un would be equivalent to about 0.15 degrees. Planetary
conjunctions which Kiang analysed, covering a span of more than
a thousand years - from the Later Han to the Yüan Dynasty -
proved less reliable, but were still consistent with this result
within a factor of 2 or so. Our own investigations tend to
confirm Kiang's conclusion (cf. Stephenson, 1971), and earlier in

this book - in Chapter 5 - we have already discussed a triple
planetary conjunction from the Later Han Dynasty (AD 182), where
it is apparent that 1 ts'un was roughly equal to 0.1 degrees.

As late as the 17th century, linear units still took the place of
angular units in the astronomy of the Far East. This is evident
from the Wu-pei-chih ("Treatise on Armament Technology") dated
AD 1628,which is quoted by Ho Peng Yoke (1966). What is particularly interesting is the discussion of planetary motions in this
work. Fast planetary motion (su-hsing) is defined as "movement
of 5 ts'un to 1 ch'ih per day", while haste (chi) is "movement
of 1 tu (i.e. degree) per day". Slow planetary motion (ch'ih-
hsing) is defined as "movement of 1 or 2 ts'un per day". The
mixing of linear and angular units here is curious. Comparison of
these rates with the mean daily motions of the planets (of the
order of 1 degree for Venus and Mars and 0.1 degree for Jupiter
and Saturn) suggests that ch'ih and tu were possibly equivalent.

In order to avoid the seemingly artificial exactness of Kiang's
result (without in any way wishing to detract from the value of
his investigation) we shall adopt the approximate conversions :

$$1 \text{ ch'ih} = 1 \text{ tu} = 1 \text{ degree}$$
$$1 \text{ ts'un} = 0.1 \text{ tu} = 0.1 \text{ degree.}$$

We learn from Professor N. Sivin of MIT (private communication)
that in his experience ts'un is not an astronomical term. Its
usage here is simply one of a great many instances in which the
Chinese used ch'ih, and its decimal divisions and subdivisions
as a way of expressing non-linear quantities to several decimal
places. However we have felt it desirable to discuss the whole
question in detail.

The term shu ("several") in text (a) normally means 3, 4 or 5
(occasionally more), so that the "approximately several inches
(ts'un)" of the record would imply some 0.3 to 0.5 degree, or
perhaps rather more.

HPP raised the question whether this last expression refers to
the position of the new star in relation to T'ien-kuan or its
apparent physical size. They state that the following interpretation of the record (a) is equally plausible: " .... (a guest
star) appeared to the south-east of T'ien-kuan (measuring)
about several inches ....". We ourselves cannot agree with this
amendment. The text reads much more easily with our original
interpretation. Additionally, although many Far Eastern accounts
of comets give an estimate of the apparent size, the word ch'ang
("length") is regularly inserted immediately before the estimate.
We have noted that this is particularly true of the Sung-shih
astronomical treatise.

If we were to adopt HPP's suggestion, this would imply that the
apparent angular size of the star was comparable with that of the
Moon (0.5 degree). Whether it was a nova or supernova, the star
would act as a point source of light. As discussed in chapter 7,
all stars appear with the unaided eye to have an apparent diameter

which is greater the brighter the star. However, even the excessively brilliant supernova of AD 1006 was only estimated to be $2\frac{1}{2}$
to 3 times the size of Venus, which itself was regarded as having
an apparent diameter one-tenth of that of the Moon. We must
return to our original interpretation - that the star appeared
several <u>ts'un</u> away from <u>T'ien-kuan</u>. Depending on just how we
understand <u>shu</u> ("several"), the star could have been anything up
to a degree away from ɜ Tau, but not significantly more than this.

Regarding the apparent magnitude of the star at brightest, we
have only the visibility in daylight of 23 days reported in (d)
to go on. The comparison with Jupiter in (f) is of little value
since, as was pointed out by Mayall and Oort (1942), at the time
the planet was an evening star setting soon after the Sun. The
two objects would not be visible simultaneously, for the new
star was only observable in the early morning before sunrise.
It might be mentioned that Venus was also on the opposite side of
the Sun from the star. At the end of the 23-day period of visibility in daylight, the new star would be some 60 degrees west
of the Sun and thus was well placed. The magnitude of Venus
varies from about -4.5 to -3.5 (except for a very brief period
around inferior conjunction). Far Eastern sightings of the planet
in daylight are fairly frequent, but are still rare enough to
be regarded as omens. This suggests a magnitude of -4 or slightly
brighter when the new star ceased to be seen by day.

Some elementary statistical considerations indicate that the
probability of the star being a nova (rather than a supernova)
is very small. If the star was indeed a nova, it must have been
very nearby. From the compilation of nova statistics collected
by Payne-Gaposchkin (1957), the absolute magnitude of a nova
at maximum varies from about -8.5 for a fast nova to -6.0 for
a slow nova. In 23 days a fast nova would fade by at least 3
magnitudes so that the apparent magnitude at maximum would be
around -7 or brighter. This would correspond to a distance of
less than 20 pc from us. The light curve of a slow nova is very
irregular, but variations of a magnitude within 23 days are quite
typical. Assuming a peak brightness of -5 leads to a distance of
some 15 pc. Let us suppose a maximum distance of 20 pc for a nova
as bright as the AD 1054.

Only observations of the brightest novae in recent times are likely
to be complete. To obtain some idea of the frequency of novae,
let us use the fact that five novae reached an apparent magnitude
brighter than +2 during the first 50 years of the present century.
From Payne-Gaposchkin (1957) the mean distance of these stars from
the Sun was 300 pc. Thus, assuming a nova occurring within 300
pc of the Sun on average once every decade, a nova would occur
within 20 pc only about once every 30,000 years, At this small
distance away, the distribution of stars is isotropic, whereas
ɜ Tau is close to the galactic equator (b = -4°). The probability
of a nova occurring so close to the Earth and in such a low galactic latitude during the last 2,000 years (the period of detailed
historical records) is of the order of 0.01.

Searches for post novae in the area around ɜ Tau have proved

negative. As discussed in chapter 1, the generally accepted
explanation of nova outbursts is mass transfer in a binary system
containing a highly evolved red star and a white dwarf of very
high surface temperature. Mayall and Oort (1942) made a survey
of the area around 3 Tau for post novae and found none. We have
decided to approach the question using the large estimated parallax
of the new star if it was in fact a nova.

At such a small distance, proper motion is likely to be large,
typically of the order of 0.2" per annum. We have checked a
detailed star catalogue containing spectra and proper motions for
stars down to magnitude +19. Of the few stars within 2 degrees
of 3 Tau brighter than magnitude +8, only two have proper motions
greater than 0.025". These are a star of spectral type F8 and
magnitude +7.5 (proper motion 0.10") and a star of type A2 and
magnitude 6.4 (proper motion 0.05"). From its intermediate spectral type, the former is certainly not a post nova, while the latter
has a spectroscopic parallax of 0.010" (Schlesinger, 1940),
implying a main sequence star.

The catalogue of trignometric parallaxes compiled by Jenkins
(1952) seems fairly complete down to magnitude +8, and contains
many stars fainter than this. Within 2 degrees of 3 Tau there
is only a single star listed closer than 50 pc. This is of spectral type G4 and magnitude +8.6. It has a well determined parallax
of 0.053" $\pm$ 0.011", indicative of a main sequence star. Such a
catalogue is likely to be particularly complete for nearby stars
since these reveal themselves by their large proper motions.

Finally, the catalogue of stellar parallaxes (many of them spectroscopic) produced by Schlesinger (1935) covers much the same range
of magnitude as Jenkins' work, but contains no extra stars nearer
than 50 pc. We might mention here that 3 Tau itself (magnitude
+3.0) is of type B3 spectrum, but very small parallax -0.14",
determined spectroscopically. The star is thus a main sequence
blue star, clearly not a post nova.

We can conclude that if the star of AD 1054 was a nova, its range
of brightness was at least 13 magnitudes. This is greater than
for all but the very fastest nova. Only two such stars are known
to have exceeded this range and both were extremely fast. Nova
Puppis (1942) reached a peak of +0.4 and the post nova is now
around +17 (cf. Payne-Gaposchkin, 1957). Nova Cygni (1975)
reached a magnitude of +1.8 (cf. Isles, 1976). At the present
time it is still declining, but a search for the pre-nova down to
magnitude +20 has so far proved unsuccessful. In 23 days Nova
Puppis declined by 5 magnitudes, and Nova Cygni by as much as 7.
The peak magnitude for the AD 1054 star would be -9 or -11 on this
basis, comparable with the Moon in brightness. That the star was
a very fast nova of this type can be ruled out by the lack of any
known sighting in Europe and the Arab Lands, and especially by
the absence of any reference to unusual brilliance in the Chinese
and Japanese sources. Bearing in the mind the low probability of
the event being a nova in any case, we feel the nova hypothesis
must be rejected. We have no hesitation in concluding that the
new star of AD 1054 was a supernova.

The remnant of the supernova of AD 1054 has long been accepted to
be the Crab Nebula (particularly since the work of Mayall and
Oort, 1942).   The Crab is an established supernova remnant.
Let us discuss the evidence in favour of and against this identification.

Our usual estimate of the maximum distance of the supernova
can be made assuming a value for peak absolute magnitude.   Unlike
novae, whose rise to maximum brightness is usually very steep,
supernovae have a rounded peak in the light curve.   This makes
discovery several days before the maximum probable.   Once more
using the 23-day period of visibility in daylight, a fair estimate
of the apparent magnitude at maximum would be -5.   Allowing
1 magnitude per kpc for interstellar absorption (Minkowski, 1968,
used the more precise figure of 1.07 for absorption in the
direction of the Crab Nebula), and a peak absolute magnitude of
-19 gives a distance of 2 to 2.5 kpc.   As discussed below, the
extrapolated date of outburst of the Crab Nebula from the rate of
expansion of the optical filaments is AD 1140+10.   This is in
fair agreement with the date of the star since acceleration of
the expansion has probably taken place (see later).   The distance
of the Crab Nebula (~2 kpc) is in excellent accord with our
rough calculation for the supernova.   Additionally, the angular
distance of the nebula from ʒ Tau (1.1 degree) fits in well with
that for the new star (about a degree at most).   Proper motions
of the nebula and ʒ Tau are negligible.

The one problem arises from the fact that the Crab Nebula lies
to the north-west of ʒ Tau, whereas according to text (a) the
star appeared to the south-east of it, i.e. in the very opposite
direction.   HPP believed that this was a serious obstacle to the
identification but they were of the opinion that sources (a) and
(c) were independent.   We have shown that this is not the case.
As the contents of the Sung-shih have probably suffered much at
the hands of the Yüan Bureau of Historiography, we need not be
overconcerned at this apparent discrepancy, particularly as three
other texts (b), (d) and (f) confirm that the star appeared in the
vicinity of ʒ Tau.   Again, the astronomer who made the original
observations may have mistaken the relative positions of the new
star and T'ien-kuan in writing up his report.   To quote a parallel
example, on a date corresponding to AD 1253 September 16, the
astronomical treatise of the Koryŏ-sa reports that Mars occulted
the south-east star of Yü-kuei.   This asterism in Cancer consists
of only four stars, each some 5 degrees apart.   Our calculations
show that in place of the south-east star (δ Can) we must read the
north-west star (η Can).   Obviously this merely indicates that
such confusion can occur.

If we do not accept this interpretation, we have to explain why
the supernova which did produce the Crab Nebula is not recorded,
and also what became of the remnant of the AD 1054 star. From the
expansion rate, the date of appearance of the star which produced
the Crab Nebula was certainly within a century of AD 1054. At the
known distance of the nebula (~2 kpc), if the star was a supernova
of Type I its peak brightness would be around magnitude -5.5, and
if of Type II around -4.   The worst possible conditions for

visibility of the star would occur if it first appeared a few
days or weeks before conjunction with the Sun. Even allowing as
much as 50 days between the initial outburst and the eventual
discovery after conjunction with the Sun (which seems excessive),
fading in the intervening time would be only some 3 magnitudes
for a Type I supernova and 1.5 magnitudes for Type II. The star
would thus be at least as bright as Jupiter when it was first seen.

The sudden appearance of such a prominent star almost exactly
on the ecliptic (the most extensively observed region of the sky)
could scarcely fail to be noticed. If no planet happened to be
near, the monthly conjunction of the Moon with the star undoubtedly
would draw attention to it. Yet there is not a single record in
oriental history which might be taken to refer to the star except
in AD 1054, and for this we find several Chinese and Japanese
descriptions.

If we are in fact dealing with two separate supernovae, we have
to explain the remarkable coincidence of two such stars appearing
within less than a century of one another, only about 1 kpc apart,
and at a mean distance of 2 kpc from the Earth. Furthermore,
these two stars appeared roughly in our line of sight. The probability of such a double event is minute - of the order of 10^{-5}.

Again, what of the remnant of the AD 1054 star? A young supernova
remnant at a distance of 2 - 2.5 kpc at most would be expected
to be detected as a radio source. However there is no known
supernova remnant within 5 degrees of ζ Tau apart from the Crab
Nebula itself, and within 20 degrees the only 3 SNRs (PKS 0607+17,
S147, IC443) are all of extreme age.

We feel that the various lines of evidence discussed above justify
the reversal of the recorded direction of the guest star of AD 1054
with respect to ζ Tau as recorded in text (a), and permit reinstate-
ment of the Crab Nebula as one of the few select remnants of
precisely known age.

HPP, by drawing attention to the problem of the relative directions
of ζ Tau and the Crab Nebula, and other supposed difficulties
which we have answered above, were no doubt expressing sincere
reservations. However, their paper is based almost entirely on
destructive criticism with very little constructive effort (apart
from the valuable identification of T'ien-kuan). Such a biased
approach is contrary to the interests of science. Ho Peng Yoke's
contribution to the field of oriental astronomy has been outstanding,
but in this particular case we feel that his efforts have been
misguided.

While still on the subject of the historical records, we might
mention that Miller (1955a and b) drew attention to two possible
rock art records of the AD 1054 supernova. He believed that these
represented a close conjunction of the Moon with the star and
calculated that such a configuration took place on the morning of
5 July in AD 1054. This is very soon after discovery in China.
Miller's suggestion was enthusiastically taken up by Brandt
et al (1975), who provided further evidence, although they pointed

out that it was not possible to prove their conclusions.
Archaeological findings indicate that the period is roughly correct
(within a century or so). Our own reaction to this fascinating
hypothesis is favourable. However, we emphasise that the pictographs tell us nothing about the star. On the other hand, Ellis
(1975) stresses that the Pueblo people "characteristically did not
record exciting events" ,and gives several examples. She believes
the pictographs to be nothing more than clan symbols. At the
moment, the question seems open, but even a satisfactory interpretation does not affect our analysis.

It should be noted that from the historical data it is not possible to decide whether the supernova was of Type I or Type II.
All we know about the light curve is that the star reached a
magnitude close to -4 around 27 July in AD 1054 and disappeared
from sight on or about 16 April in AD 1056. When the star finally
faded, it was some 40 degrees east of the Sun, and was thus well
placed for visibility, especially in view of its northerly situation. This sets a limit of about +5, or a little fainter some
21 months after it reached -4. Type I light curves have been
followed for about this length of time, but not Type II
(cf. Zwicky, 1965). No Type II supernova seems to have been
followed for more than about 150 days or so, and the form of the
light curve after this time is entirely hypothetical. A clear
distinction could be made if more data on Type II light curves
could be obtained over a longer period, and we regard this as a
matter of some urgency.

For some considerable time the AD 1054 supernova was classified
as a Type I event. As Minkowski (1971) has pointed out this classification was actually something of an historical accident.
At the time of early suggestions of an association between the
Crab Nebula and the AD 1054 outburst (for example Mayall and
Oort, 1942), similarities with the supernova in the galaxy IC4182
were emphasized. This latter supernova was later to become the
prototype of Type I supernovae, and the assignment was automatic-
ally (but without any sound basis) tagged to the supernova of
AD 1054. There is in fact ample evidence that the AD 1054 supernova
could not have been a simple Type I event (Minkowski, 1966; 1968).

An often quoted observation of Geoffrey Burbidge is that modern
astronomical investigations have been equally divided between
studies of the Crab Nebula and studies of everything else in the
Universe. While this is undoubtedly a gross exaggeration, it
does highlight the unique place of the Crab Nebula in astronomy.
No other object has stimulated so many new theories, or so readily
provided the observational means to test them. It was the first
galactic object recognised to be the remnant of a supernova; the
first radio source and the first X-ray source to be identified
with a particular object (other than the Sun); the first source
for which linear polarisation, indicative of synchrotron emission,
was observed; the first (and to-date only) remnant of an historically observed supernova for which a pulsar has been detected;
the first SNR with an optically observed pulsar. The Crab Nebula
is one of only two SNRs displaying extended radio, optical, and
X-ray emission plus a pulsar, (the other is the Vela SNR);

but despite all of these features, and intense observational and
theoretical study, it remains a unique and enigmatic object
displaying few of the common evolutionary properties of nearly all
other SNRs.

Optically the Crab Nebula must be one of the most familiar of
stellar objects. The discovery of the nebulosity has been attributed to an English amateur astronomer John Bevis in 1731. It
was 'rediscovered' in 1758 my Messier in his observations of the
passage of Halley's comet in that year, and was listed as M1
in his famous catalogue of objects that might cause confusion in
cometary studies (the first catalogue of nebulae). The common
name of the Crab Nebula seems to have been acquired about a century
later, and may have arisen from an 1844 drawing of the object by
Lord Rosse which resembled a crab— certainly the name has been
in general use from about that time. The first photograph of
the Crab Nebula was obtained by Robertson in 1892 using a 20 inch
reflector; Plate 9 shows a 1952 photograph using the Palomar
200 inch reflector.

Early spectral observations of the nebula were by Mayall (1937)
and others; these indicated that the predominant emission lines
were O II , O III , N II , S II , and the Balmer series of
hydrogen. A distinctive feature was the strength of the continuum
emission. Subsequently Minkowski (1942) detected relatively weak
He lines. He isolated all the emission lines to a system of fine
filaments, and in fact the nebula is now recognised to have two
dissimilar components; an amorphous, bluish region, plus multicoloured filaments which are predominantly green near the centre
of the nebula. The blue region is generated directly by the
synchrotron process - it is believed that the electrons are
accelerated to the speeds required to emit synchrotron radiation
in the strong magnetic field of the active pulsar. Polarization
of this component, indicative of its synchrotron origin, was first
demonstrated by Dombrovsky (1954).

The Crab Nebula remains the only SNR with optical emission
unambiguously identified as synchrotron in nature. (The optical
emission from all other SNRs is thought to result from shock-
heated interstellar material.) The synchrotron emission extends
into the far ultra-violet, and this radiation ionizes the gas
filaments which are then seen in the emission lines characteristic
of their chemical composition. The more energetic synchrotron
radiation photons are believed responsible for doubly ionizing
oxygen producing the characteristic green lines seen near the
centre of the nebula. The exact interpretation of spectral data
in terms of chemical composition remains open to some debate.
An unexpected result of Davison and Tucker (1970) is that the
abundance ratio of helium to hydrogen is a factor of from 6 to
10 larger than normal, providing evidence of nuclear transform-
ations in supernovae.

Comparison of line emission photographs of the Crab Nebula taken
ten or more years apart show that the object is expanding. As
noted earlier, the time scale of this expansion is compatible with
its origin in the AD 1054 event, with its centre within about 10"

Plate 9. The Crab Nebula.

of the pulsar.

The filamentary expansion was discovered by Duncan (1921), with
further investigation by Duncan (1939) and Baade (1942). More
recent observations have not greatly affected the interpretation
of these early results. The angular expansion in the direction of
the major axis of the nebula was determined to be 0.235"+0.008"
per year, which for the size of the nebula and assuming a uniform
expansion rate places the date of the outburst as AD 1140+10 .
In view of the certain association of the Crab Nebula with the
AD 1054 supernova, this anomaly must be interpreted as indicating
an acceleration over the history of the nebula - if constant,
this acceleration amounts to 0.0014 cm/sec^2 for the assumed distance of the nebula, with the initial expansion velocity being
~1700 km/sec along the major axis of the nebula and ~1100 km/sec
along the minor axis. Trimble (1971) has suggested that this outward acceleration could be imparted by the outward pressure of a
magnetic field and/or a relativistic particle flux with values
comparable with minimum estimates required to produce the observed
synchrotron radiation.

Fig. 8.2. A 5000 MHz map of the Crab Nebula,
from Wilson, 1971. The cross marks the pulsar.

The association of the Crab Nebula with the strong radio source
Taurus A was confirmed by the observations of Bolton and Stanley
(1949), this being the first positive identification of a discrete
cosmic radio source with an optical object. Since that time
the Crab Nebula has probably been subjected to more intense
observation than any other radio source, so that today its spectral
and structural properties are well understood. High-resolution
maps (see Fig. 8.2) show that the radio distribution is remarkably
similar to that of the optical continuum, although larger by a
factor of $1\frac{1}{2}$ to 2, depending on direction. The distribution is
approximately centred on the Crab Pulsar NP 0532, first detected
at radio wavelengths by Staelin and Reifenstein (1968). It is
essentially amorphous with no hint of the peripheral brightening
characteristic of the vast majority of radio remnants of supernovae. Many of the bright ridges of radio emission appear to be
associated with the bright optical filaments; Wilson (1971) has
suggested that this could be explained by a magnetic field enhance-
ment of 1.5 or more near the filaments.

The radio spectrum of the Crab Nebula has been thoroughly investigated. Although observations extend from 10 MHz to 250 GHz (see
Fig. 8.3) the most accurate measurements are in the range 0.1 to
10 GHz. At lower frequencies ionospheric and inter-planetary
scintillation effects make observations difficult; at higher
frequencies calibration difficulties could help to explain the
wide dispersion in flux density estimates. Within the range 0.1
to 10 GHz the spectral index α = -0.26 , somewhat flatter than
typically found for SNRs. Baldwin (1971) has investigated the
complete electromagnetic spectrum (see Fig. 8.4) and shown that
it fits a simple smooth curve over the whole observed range,
having a spectral index of -0.26 for radio wavelengths (10^7 to
10^{11} Hz), -0.9 for infrared and optical wavelengths (6 x 10^{13} to
10^{15} Hz), and -1.2 for X-ray wavelengths (2 x 10^{17} to 10^{20} Hz).
The overall spectrum may be interpreted as a synchrotron spectrum
extending from the radio to the X-ray region (Shklovsky, 1966),
requiring a permanent injection of high-energy electrons with
total energy of $\sim 10^{38}$ ergs/sec.

In contrast to optical polarisation greater than 50 percent in
parts of the nebula, radio polarisation (first detected by Mayer
et al, 1957) is less than 1 percent at frequencies below 1 GHz,
increasing slowly to about 7 percent at 10 GHz; Faraday rotation
in the shell of filaments is believed to obliterate the polarisation at low frequencies. Nevertheless, the observed degree
of polarisation is entirely consistent with a wholly synchrotron
origin for the radio emission.

At X-ray wavelengths the Crab consists of a pulsed component
(clearly associated with the pulsar) and an extended continuum
emitter. Even prior to the advent of X-ray astronomy SNRs were
proposed as likely X-ray sources due to emission from heated
swept-up interstellar material. Detection of the Crab Nebula
as a strong source of X-rays (Bowyer et al, 1964) was therefore
hardly unexpected, and represented the first positive identification
of a cosmic X-ray source. The nature of the X-ray emission was
however to provide many surprises. The extended continuum emitter

Fig. 8.3. The radio spectrum of the Crab Nebula
(from Baldwin, 1971).

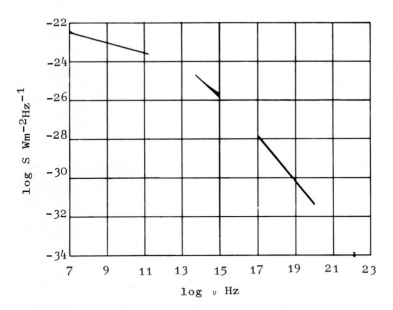

Fig. 8.4. The electromagnetic spectrum of the
Crab Nebula (from Baldwin, 1971). The thickness
of the line covers the limits of uncertainty at
any frequency.

is about 60" by 90" in extent (Hawkins et al, 1974), with a well
determined power-law spectrum indicating radiation by the synchrotron process rather than the expected thermal emission from a hot
gas. The observation by Kestenbaum et al (1971) that the X-ray
emission is polarised confirms that high-energy electrons are almost certainly responsible for the emission. In this regard the
Crab is unique, for there are at the present time no other known
SNRs for which the synchrotron process dominates the X-ray emission.
Extended soft X-ray emission from beyond the central extended
component has recently been detected; this emission may result
from the passage of the shock wave, and the significance of this
observation will be discussed in Chapter 12.

The Crab pulsar NP 0532 is the only pulsar to have been found at
optical radio, and X-ray wavelengths. Prior to the detection of
pulsed emission, the source had been identified as a compact
radio object with steep spectrum by Hewish et al (1964) and as
a 16th-magnitude star, which had been suggested to be the stellar
remnant of the supernova as in the scenario of Zwicky (1939).
The pulsar period of 0.0331 second is gradually increasing at a
rate of 3.5×10^{-8} second per revolution per day; these values
are completely compatible with the pulsar's formation about 1000
years ago with a doubling of the period in that time, and provide
yet another link between the Crab Nebula and the AD 1054 new star.
The discovery of the pulsar, and the realization that it could
provide a continuing supply of high-energy electrons, solved the
long-standing problem of explaining the large amount of energy
still being radiated by the entire nebula so long after the
outburst. The estimated loss of rotational energy of the pulsar
of 5×10^{38} ergs/sec is comparable to the total radiative energy
losses, confirming that the pulsar provides an adequate energy
source for the Nebula over the complete electromagnetic spectrum
(Woltjer, 1971).

The uniqueness of the Crab Nebula is no doubt sufficient justification for its intensive investigation in the past. However,
despite being one of the few known supernova remnants of precisely
known age, it is of no use in studying the general properties
and evolutionary behaviour of SNRs, since not one other galactic
SNR resembles it in detail. For this reason we feel that the
Crab Nebula may have dominated the attention of astronomers for
too long; we would express the hope that the rather more typical
remnants of other historical supernovae may in future enjoy
the observational attention so lavishly afforded the Crab.

Chapter 9

THE INVADER OF THE GUEST HOUSES

The guest star which appeared in AD 1181 probably aroused more
interest than any other new star which was seen before modern
times, with the exception of the brilliant supernovae of AD 1006,
1054, 1572, and 1604. The star was observed independently in
north and south China (the Kin and Sung empires) and also in
Japan. As well as being mentioned in several Japanese histories
the guest star was reported in the diaries of certain courtiers
living at the time.

We have consulted the original Japanese records in the references
cited by Kanda Shigeru (1935), as well as the original Chinese
records. Translations of these are as follows:

(a) Sung-shih ("History of the Sung Dynasty",chapter 56).
"On the day chi-szŭ in the sixth month of the eighth year of the
Ch'un-hsi reign period (AD 1181 August 6) a guest star (k'o hsing)
appeared in K'uei-hsiu and invading (fan) Ch'uan-shê until the
day kuei-yu of the first month of the following year (AD 1182
February 6), altogether 185 days; only then was it extinguished
(mieh)."

(b) Kin-shih ("History of the Kin Dynasty",chapter 20).
"On the day chia-wu in the sixth month of the 21st year of the
Ta-ting reign period (AD 1181 August 11) a guest star was seen
at (yu) Hua-kai altogether for 156 days; then it was extinguished".

(c) Meigetsuki (Diary of Fujiwara Sadaie, AD 1180-1235)
"On the day kêng-wu, the 25th day of the sixth month of the first
year of the Yōwa reign period (AD 1181 August 7) a guest star
appeared at the north near (chin) Wang-liang and guarding (shou)
Ch'uan-shê."

(d) Dainihonshi ("History of Great Japan"; written AD 1715). The
account is identical with that in the Meigetsuki, and is presumably derived directly from this earlier work.

In the following Japanese description we have omitted the date
when it is identical with that in the Meigetsuki record. We are
grateful to Dr. M.I. Scott, of the University Library, Cambridge,
for supplying the notes shown in parentheses.

(e) Gyokuyō (Diary of Fujiwara Kanezane, AD 1164-1200, courtier
of the imperial court).
"On the 28th day (commentary:kuei-yu)it was heard that since the
25th day a guest star had been present in the inner sky (commentary: beside Wang-liang), a sign of abnormality indicating that

at any moment we can expect control of the administration to be lost."

(f) Hyakurenshō (History of the imperial court from AD 967 to 1259. Written about AD 1260).
"A guest star was seen at the north pole."

(g) Azuma Kagami (History of the Kamakura military government, AD 1180-1266. In two parts: mid 13th and mid 14th century.)
"At the hour hsü (19^h to 21^{h}local time) a guest star was seen in the north-east. It was like Saturn and its colour was bluish-red and it had rays. There had been no other example since that appearing in the 3rd year of Kankō (AD 1006)".

The record appears to be incomplete. The NASA translation of the paper by Hsi Tsê-tsung and Po Shu-jen (1965) prefers "it was as large as Saturn", while Shinoda Minoru (1960) renders the passage "it was the colour of Saturn". The translation of Yang (1966) which reads "it was as big as Tsung Hsing (110,111 Herculis)", is quite erroneous.

(h) Kikki (Diary of Yoshida Tsunefusa, AD 1140-1200, a courtier of the imperial court).
"The twenty-fifth day kêng-shen of the eighth month (AD 1181 September 26). Today was the occasion for making auspicious offerings for a good harvest. A guest star had been seen in the NNE and at the present time it still has not faded away."

The star does not seem to have attracted attention in Korea. The Koryŏ-sa ("History of the Kingdom of Koryŏ") contains detailed astronomical records for this period (12th year of Myŏngjong) but the guest star is overlooked.

The oriental asterisms mentioned in the accounts (a),(b),(c) and (d) are all in the vicinity of Cassiopeia. This location is confirmed by the Azuma Kagami record (g) which states that the guest star appeared on August 7 between 19^h and 21^h (local time) in the north-east. As the Sun was then at R.A 9^h 30^m, the star (sighted in China the night before) would probably be observed soon after 19^h 30^m local time, when darkness fell. By this time Cassiopeia (R.A about 0^h) would be considerably to the east of the pole. The observation quoted in the Hyakurenshō (f)-"at the north pole"- gives only a general indication of position.

In the Azuma Kagami account (g) the guest star was stated to be the only example since the brilliant supernova of AD 1006. To what extent comparison should be made is difficult to decide owing to the interval of nearly two centuries between the observations. The "rays" described are presumably merely the protuberances which appear surrounding all bright stars owing to distortion of light within the eye. That they should be so noticeable suggests that the guest star was considerably brighter than zero magnitude. the silence of all the other accounts on the question of the brightness of the star and the absence of reference to it from Korea do not seem to support this conclusion. However it is difficult otherwise to account for the considerable interest aroused by the appearance of the star in both China and Japan.

The comparison with Saturn in the Azuma Kagami record is rather unexpected. At the time of discovery in Japan (19^h to 21^h local time) the only planet visible was Mars, fairly high in the south-east. Jupiter would become visible near midnight, while Saturn, 17 degrees West of the Sun , would only be seen for a short time before sunrise. The magnitude of the planets were about -1,-2, and +1 respectively. Perhaps the guest star (which was visible throughout the night) most nearly resembled Saturn in brightness. There is no possibility of the characters representing the various planets being confused. It is most improbable that the reference is to the similarity in colour between the guest star and the planet even though the low altitude of Saturn may have produced colours of atmospheric origin. The description "bluish-red" needs no qualification, and the observers would have been well aware that the planet is normally whitish in colour.

The guest star was discovered in Japan one day after the first observation in southern China. Possibly the delay of five days in northern China was the result of unfavourable weather cond-itions. Kepler's supernova was discovered about eight days before the maximum, during which time the brightness increased by some 3 magnitudes (Baade,1943). Assuming that the star of AD 1181 was about as bright as Saturn at discovery, it is possible that the star became significantly brighter than magnitude 0 a few days after discovery.

The discrepancy between the periods of observation recorded by the Sung and Kin astronomers may be significant. The guest star remained visible in southern China for 24 days longer than in the north. As this was the dry season in the north, the inference is that the brightness remained near the unaided eye limit between the last Kin observation and the last Sung observation. This feature is characteristic of a Type I supernova.

On the assumption that the AD 1181 guest star was a supernova of Type I, we can attempt a deduction of the magnitude at maximum. At the time of disappearance the star was approaching conjunction with the Sun, but the high declination (about +60 degrees in AD 1181) would permit two hours of observations at an altitude of more than 30 degrees each morning before nautical twilight commenced. The star is unlikely to have been observed after the magnitude fell below +5.5; it would then be just another faint star on the verge of visibility. Tycho Brahe, under more favour-able circumstances, ceased observation of the AD 1572 supernova in Cassiopeia when it became fainter than about +5.3 (Baade,1945). The Type I light curves indicate a fall of between 5 and 5.5 magnitudes in about 180 days. This suggests an apparent magnitude close to zero at maximum, in fair agreement with the other evid-ence. The magnitude corrected for interstellar absorption may have been considerably brighter than this.

The oriental asterisms Wang-liang, Hua-kai, and Ch'uan-shê, together with Ko-tao which is not mentioned in the accounts of the guest star, are adjacent to one another in Cassiopeia and

cover an area approximately 15 degrees square. Apart from the new
star of AD 1181, the only other long-duration stars from Table 3.2
which were seen near these asterisms were those of AD 1572 and AD
1592. The new star of AD 1572, which appeared beside <u>T'sê-hsing</u>
(propably κ Cas) was Tycho's supernova. The two guest stars of
AD 1592 (labelled (B) and (C) in Table 3.2) were considered earlier
in Chapters 3 and 4, where it was concluded that they were probably
slow novae.

Fig. 9.1 is a detailed chart of the Cassiopeia region for the
epoch AD 1900.0. The positions of all stars brighter than visual
magnitude +6.0 in the zone R.A 22^h30^m to 3^h30^m, dec. +55° to 75°
are shown. Numbers represent magnitudes - for the few variable
stars the range is given. Bayer stars are denoted by the appropriate Greek letter. Only a few stars have annual proper motions
in excess of about 0.1". These are β Cas, η Cas, δ Cas, and two
stars of magnitude +5.7; BD 56° 2966 (R.A 23^h8^m, dec. +57°) and
BD +63° 238 (1^h40^m, +63°). The proper motions are in order: 0.5",
1.2", 0.3", 2.1", and 0.6". The aspect of this part of the sky has
thus remained virtually unchanged since the appearance of the
guest star. The lunar mansion boundaries at AD 1181, corrected for
precession to 1900.0, are indicated by broken lines on Fig. 9.1.
The epoch 1900.0 was chosen merely for convenience since the
coordinates of the numerous stars represented on the chart were
taken from a catalogue for this epoch.

Fig. 9.1. Asterisms in the region of Cassiopeia.

The references to Ch'uan-shê, Hua-kai, and Wang-liang in the
Chinese and Japanese accounts of the guest star are quite independent. All of the star maps consulted (see chapter 2) show these
constellations, but the form of each differs to a greater or
lesser degree from one map to another, The stars forming Hua-kai
and Wang-liang can be identified with confidence, for the shape
of both of these asterisms is very well defined. However Ch'uan-
shê is very extensive and consists of faint stars. In consequence,
identification of the constituent stars is more difficult. There
are two references to this asterism in the guest star accounts -
from southern China (Sung-shih) and Japan (Meigetsuki or Dainihonshi). It is fortunate that the Chinese observation is the
more valuable, for three of the star charts used in this investigation were constructed in China at about this time. These star
charts were discussed in chapter 2.

It seems that stars brighter than mag.+5 were seldom omitted
from the oriental asterisms. Both Hua-kai and Ch'uan-shê certainly contain several stars as faint as +5.5 but none fainter than
this. As already mentioned, Hua-kai can be reconstructed reliably.
Two of the stars of this group are of magnitude +5.5, while
Yü-kuei, the 23rd lunar mansion (the trapezium in Cancer surrounding Praesepe), contains two stars of magnitude +5.5 and +5.6. The
limit for a recognisable feature seems to lie very close to
magnitude +5.5; stars fainter than this can probably be glimpsed
in favouable conditions. The Chinese astronomers, like their
western counterparts, counted only seven stars in the Pleiades
(Mao, the 18th lunar mansion). These must have been 17 Tau (mag.
+3.8), 19(+4.4), 20(+4.0), 23(+4.3), 25 (+3.0), 27(+3.8) and
either 28 Tau (+5.2) or 16 Tau (+5.4). Other stars in this cluster
brighter than +6.0 are 18 Tau (+5.6), 21 Tau (+5.9) and BD +22°
563 (+5.5). All of these stars must have been overlooked.

As well as the asterisms Ch'uan-shê, Hua-kai and Wang-liang it is
necessary to attempt reconstruction of Ko-tao which lies close
to all of these formations. On the Su Sung chart, Ko-tao actually
intersects Ch'uan-shê, a very rare occurrence. Two isolated stars,
Tsê-hsing which forms part of Wang-liang, and Fu-lu, part of
Ko-tao, are also included in this investigation.

(a) Hua-kai
According to the Chin-shu (Ho Peng Yoke, 1966), "On top of (the
star) Ta-ti (polaris) and suspended above the Imperial Throne are
the seven stars of Hua-kai ("Gilded Canopy"). Below (i.e to the
north of) Hua-kai forming its handle are the nine stars of K'ang
("Flagstaff")." The shape of this compact group is so characteristic that although the constituent stars are rather faint, it is
immediately recognizable on all the star maps we consulted. The
formation is most carefully portrayed on the Jesuit, Su Sung and
Soochow maps. The Jesuit maps show Hua-kai lying in K'uei-hsiu.
On both the Soochow and Su Sung maps the asterism is situated in
Lou, the adjacent lunar mansion, but this is obviously an error.
The stars are without doubt 32 Cas (mag.+5.3), 36 Ψ (+5.0),
40 (+5.5), 43 (+5.5), 46 (+5.0), 48 (+4.6) and 50 Cas
(+4.1). The star 40 Cas, at the north, probably forms part of the
"flagstaff" (K'ang), which extends much to the north.

(b) Wang-liang and T'sê-hsing

Named after a famous early charioteer, this group represents a chariot and a team of four horses, together with a whip. By definition, "Wang-liang, comprising five stars north of K'uei (15th lunar mansion) lies at the centre of the Milky Way (i.e at its northernmost part) and (denotes) the charioteer of the Son of Heaven. Four of these stars are known as T'ien-szŭ ("Celestial Quadriga") while the one by the side is called Wang-liang .... The single star called T'sê-hsing ("Whip Star"), the whip used by the Wang-liang.....is found by the side of Wang-liang" (Ho Peng Yoke, 1966). On all the maps we used β Cas represents Wang-liang himself. The main charts seem agreed that the remaining stars (representing the team & whip) are(in order of R.A), κ , α , η , ν (regarding ν' and ν² as one star) and γ Cas. Whereas on the Jesuit planispheres γ Cas is named as T'sê-hsing, the other charts point to κ Cas at the north. Support for κ Cas as the whip comes from the records of the supernova of AD 1572. Both Chinese and Korean records describe this star as being T'sê-hsing. The corresponding radio source (3C 10) is only 1.2 degrees north-west of κ Cas. The distance from γ Cas is 5 degrees. We have adopted κ Cas as the "whip".

(c) Ko-tao and Fu-lu

These asterisms are described in the Chin-shu as follows. "Ko-tao ("Hanging Gallery"), consisting of six stars in front of Wang-liang, like a flying path (across mountains), stretches from the "Purple Palace" (the circle of perpetual visibility) to the Milky Way.... The star Fu-lu ("Auxiliary Road") south of Ko-tao (represents) the alternative side-routes."

Both Ko-tao and Fu-lu show considerable diversity of form on the various charts. On the Jesuit maps the stars of Ko-tao are certainly ι , ε , χ , φ , θ , and either ξ or ν Cas. Fu-lu is η Cas. These last three stars lie to the south of the region covered by Fig. 9.1. It is clear from the relative positions of ι Cas, ε Cas and BD +63°265 (the +5.2 mag. star lying between ι and ε Cas) that the latter star forms part of Ch'uan-shê. On the Su Sung map the five stars of Ko-tao are ι , ε , χ , φ and θ Cas, while Fu-lu is δ Cas. Here Ch'uan-shê intersects Ko-tao between ι and ε Cas. Ch'uan-shê is very idealized, but it is clear from the position of the K'uei-Lou boundary that BD +63°265 is one of the stars of this asterism. The Korean planisphere appears to confirm the Jesuit identification of Ko-tao but replaces η Cas by χ Cas as Fu-lu.

The remaining representations are very idealized. The Soochow chart points to δ Cas as Fu-lu, but Ko-tao is very distorted, most probably containing θ , φ , χ and ε Cas. On the Tunhuang map Ko-tao and Fu-lu form a single asterism, presumably, from its shape, consisting of ι , ε , χ , δ , φ and θ Cas.

The formation shown in Fig. 9.1 seems to be the most probable one, but obviously there is considerable doubt about Fu-lu. However, we can be fairly confident that within the boundary of Fig. 9.1 all of the stars ι , ε , χ , δ , φ and θ Cas, and only these, are members of the two asterisms.

(d) Ch'uan-shê

From the Chin-shu; "Ch'uan-shê ("Inns" or "Guest Houses"), comprising nine stars and situated above (i.e to the south of) Hua-kai near the Milky Way, represents the guest-houses......" (Ho Peng Yoke, 1966). This undistinguished constellation is very extensive. All of the maps (except those in the Hsing-ching) show Ch'uan-shê situated immediately to the south of Hua-kai. On the Jesuit, Soochow and Korean maps the asterism is represented as lying to the north of, and entirely outside, the Milky Way - in accord with the Chin-shu description.

The Jesuit representations of Ch'uan-shê show only five stars. Measurement of position indicates that four of the stars are 10 Cas (mag. +5.5), BD +63°99 (+5.5), BD +63°149 (+5.5) and BD +63°265(+5.2). The fifth is probably 32 Cas (+5.5) displaced so that it is in line with the others. This formation is shown in Fig. 9.1. On both the Soochow and Su Sung maps Ch'uan-shê ranges much farther to the east, but the five easternmost stars appear to be identical with those on the Jesuit planispheres. The gap between BD 63°265 and the first of the eastern stars, probably BD 65°340 (mag.+4.8), is clearly shown in each case. On the Soochow chart the group has a zigzag pattern, seemingly an attempt to portray the true configuration.

All of these stars are faint but it is difficult to accept alternative identifications; there are no other suitable stars between Hua-kai and the Milky Way. The star κ Cas, identified above as T'sê-hsing is certainly not part of Ch'uan-shê for it lies well within the northern edge of the Milky Way. The occurrence of two independent references to Ch'uan-shê in the accounts of the guest star suggest that it was a well-known formation. This is apparent from the frequent allusions to the asterism in cometary records (e.g in AD 451, 626, 1337 and 1593).

The expression fan ("invade", "offend", "trespass against", etc.) which occurs in the Sung-shih account of the guest star is very frequently used in far eastern records to describe conjunctions of the Moon and planets with one another and with stars. The term is defined in chapter 157 of the Wu-pei-chih ("Treatise on Armament Technology" - dated AD 1628) in several ways. These are mainly descriptive, but a numerical definition is as follows: "When (a celestial body) comes within 7 ts'un of (another) such that their rays extend towards each other" (Ho Peng Yoke, 1966). The Chin-shu underlines the astrological significance of the expression: "When the planets are within a distance of seven inches (ts'un) from each other the omen will assuredly be fulfilled." This apparent confusion of linear and angular units was discussed in chapter 8 where it was concluded that ch'ih (of 10 ts'un) and tu (degrees, numbering 365.25 to a circle) are synonymous or at least roughly equivalent.

An investigation of numerous Chinese and Korean observations of the Moon and planets invading one another, or stars, reveals that fan was usually used when the least separation was about half a degree, and only very rarely for a minimum approach of more than

a degree. This seems to confirm the figure of 0.7 degrees from
the Wu-pei-chih. When an occultation occurred the more precise
terms yen (conceal), shih (eclipse), or ju (enter) were used.
Occasionally very close approaches of planets to one another, or
to stars, were described in these ways. An unexpected outcome
of the investigation is the evidence that although the civil day
began at midnight, the astronomers of both countries adopted an
astronomical day beginning at sunrise - normally a readily defined moment.

The use of the expression fan does not necessarily imply motion.
In the context of the guest star record the reference is to a
star, not previously seen, suddenly appearing close to one of the
permanent members of an asterism. In order to examine the reliability of the term as it applied to Sung observations we have
analysed a series of planetary conjunctions which were observed
between AD 1167 and 1200. These are recorded in the Sung-shih
and cover four reign periods. In each case it is stated that one
planet invaded another. The result of this analysis are shown in
Table 9.1.

Ref. No.	Recorded Date		Planets	Actual Date		Time	Distance	Elong.
1	1167 Dec.	24	V,J	Dec.	24	20^h	$0.57°$	53°E
2	1168 Apr.	14	M,S	Apr.	18	2	0.65	78 W
3	1170 Sep.	9	M,J	Sep.	13	5	0.09	106 W
4	1172 Jun.	18	V,J	Jun.	18	20	0.97	17 W
5	1173 May	11	M,J	May	12	21	0.61	71 E
6	1173 Sep.	5	V,J	Sep.	6	5	0.32	18 W
7	1175 Oct.	25	V,M	Oct.	25	18	0.41	32 E
8	1181 Aug.	14	V,S	Aug.	15	4	0.16	23 W
9	1184 Aug.	31	V,J	Sep.	1	4	0.69	35 W
10	1187 Nov.	14	V,S	Nov.	15	6	0.47	36 W
11	1189 May	22	V,M	Jun.	2	4	1.52	40 W
12	1191 Dec.	31	V,J	Dec.	31	18	0.38	45 E
13	1195 Oct.	11	V,M	Oct.	12	5	0.30	41 W
14	1198 Nov.	18	V,J	Nov.	19	6	1.06	46 W
15	1199 Dec.	2	M,J	Dec.	5	6	0.56	34 W
16	1200 Jan.	1	V,S	Jan.	1	18	0.95	45 E
17	1200 May	22	M,S	May	25	4	0.26	83 W

Table 9.1.
Planetary conjunctions recorded in the Sung-shih.

Column 2 of Table 9.1 gives the recorded date converted to the
Julian Calendar using the tables of Hsüeh Chung-san and Ou-yang I
(1956). Column 3 gives the initial letter of one of the four
planets Venus, Mars, Jupiter, and Saturn (observations of Mercury
are extremely rare). Columns 4 and 5 give the calculated date and
local time (to the nearest hour) of closest visible approach. The
time chosen is the most favourable time for observation of the
conjunction, and is within 12 hours of the time of closest
approach. Positions of Venus and Mars were computed using the
orbital elements given by Newcomb (1895). For each planet an
unperturbed orbit was assumed. This treatment is quite adequate
for the present purpose. Positions of the slower moving outer

planets were found by interpolating in Tuckerman's Tables (1964).
Column 6 gives the separation of the two planets at the time given
in column 5. Column 7 gives the elongation of the planets east
or west of the Sun.

Of the 17 observations, the recorded dates of 11 are exact (1,4,6,
7,8,9,10,12,13,14 and 16). For observation No. 11 the date is 10
days in error. This implies that the cyclical day i-wei of the
record should be replaced by i-szŭ although the characters wei
and szŭ are quite dissimilar. This observation seems to be very
unreliable, for the closest visible approach (1.52 degrees) is
about 0.5 degrees larger than any other in Table 9.1. For the
remaining 5 observations (2,3,4,15 and 17) the Sung-shih records
a date differing by up to 3 days from the calculated date. If the
record gives the true date of observation, the amended separations
(in degrees) are respectively 1.31,0.88,0.78,1.06 and 1.35. It
seems quite possible that cloud interfered with some of the
observations. However, the average of these five results is about
2.5 times that of the corresponding figures in Table 9.1 and
twice that of the remaining quantities in the table (excluding
No. 11). An error in the recorded date is thus a more probable
assumption. This is a common occurrence in lunar records, for
which the intended date can be established with certainty owing
to the rapid motion of the Moon. The Sung planetary observations
thus indicate a maximum separation of about a degree for the term
fan to be applicable, in accord with the limit deduced from other
oriental observations.

As discussed in chapter 11, an interesting Korean observation of
the planet Venus invading the guest star of AD 1604 (Kepler's
supernova) is recorded in the Sŏnjo Sillok on a date which
corresponds to AD 1605 January 20 (Gregorian). Calculation shows
that before dawn on January 21 (astronomical date January 20)
in AD 1605, Venus and the star would be seen only 0.51 degrees
apart. It thus seems probable that throughout the far east a
sighting tube was used in order to decide when the expression
fan was appropriate.

Within the right ascension range 22^h30^m to 3^h30^m, and north of
declination +55°, the Clark and Caswell (1976) catalogue lists
six SNRs; Cas A, CTB1, CTA1, 3C 10, 3C 58, and HB3. CTB1, CTA1,
and HB3 are all of large angular size and very low surface
brightness indicative of their being of extreme age. 3C 10 is
the well-established remnant of Tycho's supernova (chapter 10).
As already mentioned in chapter 4, Cas A is undoubtedly the
remnant of a recent (≈200 years), nearby, but undetected, supernova. 3C 58 thus remains as the only "young" SNR in the region
of interest which could be the remnant of the AD 1181 event.

The SNR 3C 58 fits in a general sense the various oriental
descriptions of the position of the guest star of AD 1181 - "in
K'uei-hsiu and invading the stars of Ch'uan-shê" (Sung-shih):
"at Hua-kai" (Kin-shih): "near Wang-liang and guarding Ch'uan-
shê" (Meigetsuki and Dainihonshi). As shown in Fig. 9.1,3C 58
lies just within K'uei-hsiu. It is 8 degrees from Wang-liang
but this constellation contains some of the brightest stars

in Cassiopeia. The proximity to <u>Hua-kai</u> (4 degrees) is in accord
with the Kin account.

The mention of the expression <u>fan</u> ("invading") in the <u>Sung-shih</u>
record is critical. As already seen this is an important astrological term implying a maximum distance of about one degree
from a celestial body. 3C 58 lies only 0.68 degrees from BD
63°265 (mag.+5.2) and 0.53 degrees from 53 Cas (mag.+5.6) in
<u>Ch'uan-shê</u> (the "Guest Houses").

Bearing in mind the ample opportunity for observation of the
guest star, 3C 58 fits all of the evidence from the purely astronomical point of view. The only difficulty in accepting this
identification is the lack of reference to <u>Ko-tao</u>. However, this
asterism is only rarely mentioned in accounts of both comets
and novae (in contrast to <u>Ch'uan-shê</u>). Possibly the explanation
is astrological.

The true nature of the source 3C 58 has been the subject of
considerable debate in recent years. On the basis of its angular
extent, low galactic latitude, and flat spectrum (α =-0.1) it
was originally classified as an H II region. Its non-thermal
nature was eventually revealed from polarisation measurements
by Weiler and Seielstad (1971), and Kundu and Velusamy (1972).
There can now be no doubting a synchrotron origin for the continuum radio emission from the source. Nevertheless high-resolution maps (see Fig. 9.2) fail to show the peripheral brightening characteristic of a supernova remnant. The amorphous structure of the source and unusually flat spectrum immediately liken
3C 58 to the Crab Nebula radio source; here the similarity ends,
since no central pulsar or optical nebulosity has yet been detected for 3C 58.

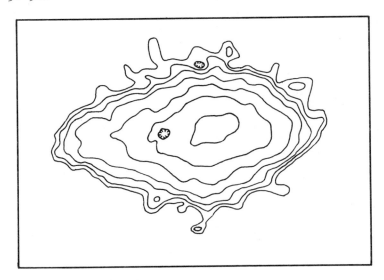

Fig. 9.2. A 15 GHz map of 3C 58 (from Green et al, 1975)

H I absorption measurements by Goss et al (1973), and Williams
(1973) place 3C 58 at or beyond the distance of the outer galactic
spiral arm at about 8 kpc. Despite this distance, the large z-
value for the source (~150 pc) and consequent lower-than-average
interstellar absorption would still allow an observed magnitude
for the supernova of about 0 (see earlier). Since the H I absorption results represent a minimum distance, there remains the
possibility that 3C 58 could be extragalactic; indeed there are
sources with similar amorphous structure (for example 3C 386)
which have been positively identified as being extragalactic.
However the flat spectrum of 3C 58 is even less characteristic
of an extragalactic source than of an SNR. Despite these mis-
givings, the majority of investigators now favour an SNR interpretation for the source.

The AD 1181 supernova presents another example where the possible
association of an SNR with the new star depends critically on the
exact interpretation of a single word in the historical records. In
this case we must rely on the astronomical precision which we
have argued is implied by the use of fan (invading); then the
excellent positional agreement between the radio source 3C 58
(R.A $2^h01^m43^s$, dec.$+64°36.5'$) and the star BD $63°265$ (R.A 1^h55^m
55^s, dec.$+64°22.8'$, with negligible proper motion) in Ch'uan-shê
is convincing evidence for 3C 58 being the SNR. In fact there is
no other remnant in the vicinity that could possibly correspond,
although because of some remaining doubts as to the true nature
of the radio source we feel we must limit the classification in
this case to 3C 58 being the probable remnant of the supernova
of AD 1181.

The Sung dynasty astronomers hold the distinction of having
recorded three supernovae - those of AD 1006, 1054, and 1181. It
was to be almost 4 centuries before another supernova was sighted
from Earth, by which time the standard of European observational
astronomy had far surpassed that of the Far East.

Chapter 10

THE NEW STAR OF TYCHO BRAHE

During the almost four centuries which were to elapse before the next supernova appeared, very little progress was made in Far Eastern astronomy. The new star which appeared in AD 1572 was observed both in China and Korea (there is no known extant record from Japan), but the reports from these countries are in no way superior to those of new stars seen a thousand years earlier. However, in Europe the Renaissance was now well under way. We today are indeed fortunate that an observer of the calibre of Tycho Brahe was flourishing when the star appeared. The careful records which this eminent Danish astronomer (who lived between AD 1546 and 1601, and was not quite 26 when the new star appeared) has left us prove beyond any reasonable doubt that the star was a supernova of Type I, and allow us to establish with certainty that its remnant is the powerful non-thermal galactic radio source G120.1+1.4 (3C10).

The star appeared at a very critical time in the development of astronomical thought in Europe. Copernicus' heliocentric theory of the Universe, published in his De Revolutionibus Orbium Coelestium (AD 1543), was beginning to attract widespread attention. In this work Copernicus attacked the long-held doctrine of Aristotle and Ptolemy that the Earth was at rest at the centre of the Universe. This had formed the basis of medieval European cosmology. It was one of the basic tenets of Aristotle's concept of the Universe that the eighth sphere of the fixed stars (the first seven spheres were taken up by the Moon, Sun and five known planets) was immutable. Hence Tycho Brahe, who could not conceal his amazement at the sudden appearance of the new star (which was about as bright as Venus), was at great pains to determine whether it belonged to the eighth sphere. In a preliminary report entitled De Nova Stella and published in AD 1573, Tycho discussed his measurements of the position of the star in relation to the principal stars of Cassiopeia (within which constellation the new star appeared). He concluded: "That it is neither in the orbit of Saturn, however, nor in that of Jupiter, nor in that of Mars, nor in that of any one of the other planets, is hence evident, since after the lapse of several months it has not advanced by its own motion a single minute from that place in which I first saw it; which it must have done if it were in some planetary orbit..... Hence this new star is located neither in the region of the Element, below the Moon, nor in the orbits of the seven wandering stars but in the eighth sphere, among the other fixed stars". Tycho's measurements allow us to determine the position of the new star at the present day with remarkable accuracy - to within a very few minutes of arc.

Tycho's original discussion of the new star is very brief. However, in his Astronomiae Instauratae Progymnasmata, which was
published in AD 1602, shortly after his death, he gives a very
detailed investigation, including numerous observations made by
others. In this work the De Nova Stella is also reprinted,with
minor omissions, as part of chapter 8.

Coming now to the star itself, it could have scarcely been better
placed for observation in Europe and the Far East in view of its
location in Cassiopeia. North of about latitude 35°N, this constellation is circumpolar (i.e it never sets), and in early November, when first discovered, the star would be almost in the zenith
in the evening. An excellent and objective account of the discovery in Europe is given by Dreyer (1890). From this it seems certain that the star appeared some time between November 3 and November 6 (in AD 1572). Dreyer rejects a number of spurious claims,
several of which are clearly self contradictory, that the star
was sighted in October. Most of these are quoted in the Progymnasmata, but perhaps the most fascinating story is contained in a
French pamphlet, unknown to Tycho Brahe, and cited by Dreyer.
This little work, published in 1590, and entitled La nouvelle
Estoille apparue sur tous les Climats du Monde: Et de ses effects,
claims that the star was seen in the month of October by Spanish
shepherds "keeping watch over their flocks" (cf St. Luke, ii,8).

A valuable negative sighting in early November is given by Jerome
Mugnoz (Muñoz), an eminent mathematician and Professor of Hebrew
in the University of Valencia (Spain). The original Spanish is
quoted by Hellman (1960). This account runs as follows: "I am
certain that on the second day of November 1572 there was not
this comet in the sky, because on purpose more than an hour after
6 in the afternoon, I showed in Hontinente (Onteniente) to many
people publicly how to identify the stars, and there were many
shepherds very experienced in them (the stars), who informed me
on the 18th day in the morning a new star had appeared". Mugnoz
later measured the position of the new star in relation to three
of the stars in Cassiopeia with fairly high precision (see below)
so that his testimony is probably reliable. Mugnoz was not alone
in describing the new star as a comet . Possibly he was being
cautious. His reference to the shepherds is somewhat obscure, for
by the 11th of the month, when Tycho Brahe first saw the star,
it was already about as bright as Venus. It seems best to trust
only the first hand experience of Mugnoz himself.

There does not seem to have been any positive sighting of the new
star until November 6, when Wolfgang Schüler of Wittenberg discovered it at about 6 a.m (Progymnasmata, chapter 9). Like Mugnoz,
Schuler described it as a comet. There is the possibility that
Francesco Maurolyco of Messina may have preceded Schuler. Hellman (1960) has discussed in detail a manuscript copy of Maurolyco's
account of the star, which states that "Maurolycus, abbot of
Messina, wrote this hastily, subject to the judgement of the more
wise. 6 November 1572". This suggests that Maurolyco may have
seen the star earlier than November 6.

By November 7 the star had been sighted independently by two
others.Tycho Brahe himself did not see it until the evening of
November 11. Bad weather had delayed the discovery of the star
by this assiduous observer. He tells us: "Although several
people had preceded me (in seeing it), the air in our part of
the world had not been clear enough" (Prog., chapter 3).

Valuable supporting evidence for a date of first appearance in
early November is given by the court astronomers of China. The
star was first seen in China on a date corresponding to November
8. Unfortunately, only the month of discovery is reported in
Korea, but this corresponds to the period November 6 to December
4.

While on the subject of the discovery of the new star, it is
worthwhile commenting on the attitude to it in Europe. In
chapter 3 of his Progymnasmata, Tycho Brahe gives a graphic
account of his own feelings. When the star first appeared, Tycho
was staying with his uncle at the monastery of Herritzwadt in
Denmark. He writes :
"When on the above mentioned day (November 11), a little before
dinner.....I was returning to that house, and during my walk
contemplating the sky here and there since the clearer sky
seemed to be just what could be wished for in order to continue
observations after dinner, behold, directly overhead, a certain
strange star was suddenly seen, flashing its light with a radiant
gleam and it struck my eyes. Amazed, and as if astonished and
stupified, I stood still, gazing for a certain length of time
with my eyes fixed intently upon it and noticing that same star
placed close to the stars which antiquity attributed to Cassiopeia. When I had satisfied myself that no star of that kind had
ever shone forth before, I was led into such perplexity by the
unbelievability of the thing that I began to doubt the faith of
my own eyes, and so, turning to the servants who were accompanying me, I asked them whether they too could see a certain extremely bright star when I pointed out the place directly overhead.
They immediately replied with one voice that they saw it completely and that it was extremely bright. But despite their affirm-
ation, still being doubtful on account of the novelty of the
thing, I enquired of some country people who by chance were
travelling past in carriages whether they could see a certain
star in the height. Indeed these people shouted out that they
saw that huge star, which had never been noticed so high up. And
at length, having confirmed that my vision was not deceiving me,
but in fact that an unusual star existed there, beyond all type,
and marvelling that the sky had brought forth a certain new
phenomenon to be compared with the other stars, immediately I
got ready my instrument. I began to measure its situation and
distance from the neighbouring stars of Cassiopeia, and to note
extremely diligently those things which were visible to the eye
concerning its apparent size, form, colour and other aspects".

The general reaction in Europe to the appearance of the star
was one of extreme conservatism.Michael Maestlin, later to become
tutor to Johannes Kepler, remarked: "Various judgements and
opinions were heard about it from people who are not illiterate.

Some conceded that it was natural and permanent, not new.......
Others indeed believed that it was Vega, others Capella, others
Arcturus ; but others (said) Venus, others Saturn, others some
other star or planet" (Prog., chapter 8). The well known astronomer Cornelius Gemma did not begin to observe the position of
the star until November 26, since he thought it idle talk when
he first heard of a new star (Prog., chapter 8). Tycho himself
wrote that he later heard that in Germany country people first
called attention of astronomers to the phenomenon (Prog.,
chapter 3).

There could well be truth in this last remark, for unlettered
people would be relatively free from preconceived ideas. Certainly if a star as bright as Venus could be overlooked by much
of the astronomical world as late as AD 1572, there should be
no difficulty in understanding the complete lack of European
references to the supernova of AD 1054. Here, the vast majority
of works which could have mentioned the star are monastic
chronicles, the writers of which might be expected to hold very
orthodox views.

Before discussing the European measurements of the position and
brightness of the star, let us take a look at the Far Eastern
observations.

The most detailed account is to be found in the Ming-shih-lu, in
the annals of Emperor Shên-tsung (the Shên-tsung-shih-lu). We
read: "(6th year of the Lung-ch'ing reign period, 10th month),
(day) hsin-wei.....Previously in the 10th month on the 3rd day
ping-ch'en at night a guest star was seen at the north-east; it
was like a crossbow pellet. It appeared beside Ko-tao in the
degrees of (Tung-) pi lunar mansion. It gradually became fainter.
It emitted light in the form of pointed rays. After the 19th day
jen-shen at night the said star was orange (reddish-yellow) in
colour. It was as large as a lamp and the pointed rays of light
came out in all directions........It was seen before sunset......
At the time the Emperor saw it in his palace. He was alarmed and
afraid, and at night he prayed in the open air on the Vermillion
Steps......Editorial note: The star gradually diminished in
brightness only in the 2nd month of the 1st year of the Wan-li
reign period. When we come to the 4th month of the 2nd year it
finally disappeared".

The first part of the above account is retrospective, for jen-shen
is the day after hsin-wei, the date of the previous entry. Evidently the writer of the above account was more interested in the
Emperor's reaction (presumably he first saw the star on the day
jen-shen). We have omitted two interpretations of the star in the
text since they disturb the flow of the account. The first states
that the star was a "rayed star" (po-hsing), and second that such
stars "are also seen in the daytime" in agreement with the observation.

The date of discovery of the new star corresponds to AD 1572
November 8, and the subsequent date when the colour was orange
and the Emperor saw the star is equivalent to AD 1572 November

24. The month when the star showed a noticeable diminution in brightness corresponds to AD 1573 March 1 to 31, and the month of final disappearance to AD 1574 April 21 to May 19.

In the astronomical treatise of the Ming-shih, the obvious place to expect a detailed account of the star, we find only a brief undated mention (chapter 25): "There are also some (stars) which did not exist in antiquity but which exist now. Beside Tsê-hsing there is a guest star which newly appeared during the first year of the Wan-li reign period. At first it was large, but now it is small".

There is no mention of the star in the Ming-shih annals, but in general we have found the imperial annals of the various dynastic histories to be unproductive as far as astronomical records are concerned. However, Ho Peng Yoke (1962) discovered a brief mention of the star in the Ming-shih-kao, the draft version of the Ming-shih. Here the guest star is described as a 'broom star' (hui-hsing) but this error was corrected in the astronomical treatise of the Ming-shih. We read in the Ming-shih-kao : "(6th year of the Lung-ch'ing reign period), winter, 10th month (day) ping-ch'en. A broom star appeared in the north-east. It disappeared only during the 4th month of the 2nd year of the Wan-li reign period."

The single Korean record is also very brief. In the Sŏnjo Sillok ("Annals of the reign of King Sŏnjo") we find the following account of the star: "(5th year of Sŏnjo) 10th month. A guest star was seen at (yu) Tsê-hsing. It was larger than Venus."

If all we had to go on were the oriental records of the star, it would be immediately apparent from the lengthy duration (some 18 months) that the star was not a comet, and for just this same reason a supernova interpretation would be more likely than a nova. Additionally, the Ming-shih and the Sŏnjo Sillok independently mention the proximity of the star to Tsê-hsing. As was shown in Chapter 9, this star is identical with κ Cas.. This location in confirmed by a star map produced by Jesuits in Peking around AD 1600, and preserved in Bologna (see chapter 2). Here the star is shown about 1 degree to the west of a star whose position corresponds with that of κ Cas.

However, the positional measurement of Tycho Brahe are fully two orders of magnitude better than those of the oriental astronomers, and further, as Baade showed in 1945, it is possible to draw a remarkably accurate light curve from Tycho's comparisons with nearby stars.

Considering first the light variation of the star, the negative sighting by Mugnoz is quite valuable. Cassiopeia is one of the most familiar of all constellations since the principal stars form a distinct "W". The magnitudes of these stars lie in the range from about +2.5 to +3.5. As is evident from Fig. 10.1, the location of the new star was such that unless it was very faint it could scarcely have failed to attract attention if someone was looking closely at the constellations. We might thus reasonably

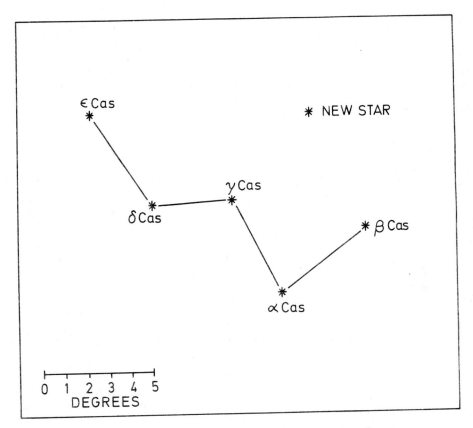

Fig. 10.1. The position of the new star with respect to the "W" of Cassiopeia.

conclude that on the evening of November 2, when Mugnoz was giving his demonstration, the new star had not reached magnitude +3.

Several observers, including Tycho Brahe, made estimates of the brightness of the new star around maximum brilliancy, but unfortunately it is not possible to establish the date of maximum with any accuracy. Tycho (<u>Prog</u>., chapter 3) estimated that when he first saw the star (November 11) it emulated Venus when the planet was at maximum brightness (which was virtually the case then). He added that it remained at nearly this same brightness throughout almost the whole of November. He further remarked that when the sky was clear many keen sighted people could see it in daylight, even at midday, while at night the star often shone through thick clouds which blotted out the other stars. However, several observers felt that the new star never quite rivalled Venus. Baade (1945) in his pioneering investigation of the light curve of the star gave considerable weight to three

other estimates made around maximum. The observers were Caspar
Peucer and John Praetorius, both of Wittenberg, and Michael
Maestlin of Tubingen. Peucer noted that the new star was
"brighter than all the planets and star with the exception of
Venus" (Prog., chapter 8). According to Praetorius, it was
"larger and brighter than Jupiter, but easily fainter than Venus"
(Prog., chapter 9). Maestlin estimated that it "surpassed Jupiter,
and almost Venus" (Prog., chapter 8).

From Baade (1943), in November of 1572 Venus was near maximum
brightness, (magnitude -4.3), and approaching greatest western
elongation from the Sun after inferior conjunction. The planet
was thus a morning star, and since the new star was circumpolar,
direct comparison could be made in the early morning hours.
Combining the various estimates, Baade deduced an apparent magnitude of the star at maximum as -4.0±0.3. This is in fair agreement with the Korean observation that the star was "larger than
Venus" and the Chinese sighting in daylight.

With regard to the date of maximum, we can do no better than
follow Baade in assuming a date near the middle of November.
Only Tycho Brahe seems to have continued to estimate the brightness of the star after maximum, but fortunately he did this
month by month. His estimates, as given in chapter 3 of the
Progymnasmata, are summarised in Table 10.1. It is perhaps to be
regretted that Tycho did not express dates more accurately than
to the nearest month, but possibly he felt that the decline was
so slow that this was justified. What is particularly important
is that he does not hint at any irregularities in the light
variation - there seems to have been a continuous decline after
maximum.

Date	Brightness Estimate
AD 1572 Dec.	About as bright as Jupiter.
1573 Jan.	A little fainter than Jupiter; much brighter than the stars of 1st mag.
1573 Feb.-Mar.	Equal to brighter stars of 1st mag.
1573 Apr.-May	Equal to stars of 2nd mag.
1573 Jul.-Aug.	3rd mag. Similar to brighter stars of Cassiopeia which are of 3rd mag. (i.e. $\alpha, \beta, \gamma, \delta$ Cas.).
1573 Oct.-Nov.	Equal to stars of 4th mag. In Nov. very similar to κ Cas.
1573 Dec.-1574 Jan.	Scarcely exceeded stars of 5th mag.
1574 Feb.	Reached 6th and faintest mag.
1574 Mar.	So faint that no longer visible.

Table 10.1. Tycho Brahe's estimates of the
brightness of the new star after maximum.

It is interesting to note that the Chinese astronomers, who continued observing the star until April - May of AD 1574, were able to follow it for a month longer than Tycho. This seems to prove the diligence of the oriental astronomers as straight-forward observers, even if the accuracy of their positional measurements was far inferior to that of the Europeans at this time.

In the above Table 10.1, there are no references to June and September in AD 1573. However, Tycho was observing the star in both months. In each case he tells us that the brightness was declining, but it was not until the following month that the star could be placed in a particular magnitude range. Probably he was making comparisons with stars of a given magnitude listed in the catalogue of Ptolemy's Almagest, and during the two months in question he could not find a suitable comparison object. How-ever, there is no doubt that he maintained a continuous watch on the new star.

For the purpose of reducing Tycho's data, Baade assumed a mean date for each comparison. He was particularly careful to check the precise magnitudes of stars in each of the classes 1 to 6 adopted by Ptolemy, and presumably followed by Tycho. It would be difficult to improve on Baade's investigation. His results for the variation in brightness of the new star are summarised in Table 10.2. A date of maximum of AD 1572 November 15 is assumed.

Days after maximum	Visual magnitude
0	-4.0
30	-2.4
61	-1.4
107	+0.3
167	+1.6
259	+2.5
351	+4.0
365	+4.2
412	+4.7
457	+5.3
485	....

Table 10.2.   The varying brightness of the new star of AD 1572.

The magnitudes derived by Baade and quoted in Table 10.2 are on the Harvard Revised Photometric scale. Obviously, as the principal objective is to produce a light curve, the precise magnitude scale is unimportant. To the above results may be added the estimate made earlier from the negative sighting by Mugnoz that the star was probably fainter than +3 on November 2 (13 days before the assumed maximum).

The light curve drawn from the data in Table 10.2 is shown in Fig. 10.2. This is taken direct from Baade's paper with the

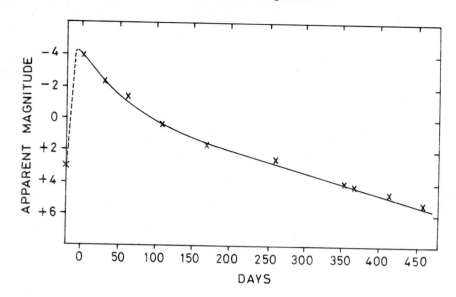

Fig. 10.2. The light curve of the AD 1572 supernova.

addition of the single pre-maximum point. Comparisons with the
composite Type I and Type II light curves shown in Figs. 1.6 and
1.7 (chapter 1) make it evident that the star was a supernova of
Type I. The light curve is clearly not that of a nova. Undoubtedly
Tycho's estimates individually were crude by modern standards,
but the resultant light curve has all the characteristics of a
Type I supernova - an initial fairly rapid decline followed by
a linear descent at about 1 mag. in 50 days.

As the light curve of the new star presents conclusive evidence
that it was a supernova rather than a nova, we should expect to
find a powerful discrete radio source as its remnant. Tycho Brahe
made a far more systematic determination of the location of the
star than any of his rivals, measuring its angular distance from
the principal stars of Cassiopeia and also Polaris to the nearest
minute of arc. From Tycho's measurements, Argelander (1864) and
Böhme (1937) deduced position which were in excellent accord.
Argelander used Tycho's measurements in relation to $\alpha, \beta, \gamma, \delta, \epsilon, 3,$
η, κ Cas and obtained a result for the epoch 1950.0 of
R.A. $0^h22^m0.2^s$, dec. $+63°52'12''$. Böhme derived two solutions,
depending upon whether he excluded η Cas or included it along
with ι Cas. The first of these (for the same epoch) is R.A.
$0^h22^m1.1^s$, dec. $+63°52'1''$, and the second $0^h22^m4.6^s$, $63°52'30''$.
Baade attributed the "unexpectedly large" difference between
Böhme's two solutions to an error in Tycho's measurement of the
distance of the new star from ι Cas. By omitting this measurement
he obtained a result identical with that of Argelander, and
estimated the mean error in the position of the star as only ±23".
Baade made a search of the area down to mag. +19 for a blue star
which might be the stellar remnant, but found nothing.

Radio emission from the remnant of Tycho's new star was first
reported by Hanbury-Brown and Hazard (1952), using a large
meridian-transit radiotelescope at the Jodrell Bank Observatory.
This discovery, unlike Bolton's chance discovery of radio emission
from the Crab Nebula (see chapter 8), was the result of a planned
search and represented only the second positive identification
of a galactic supernova remnant. The coordinates of the radio
source were given by Hanbury-Brown and Hazard as R.A. (1950)
$0^h21^m49^s\pm2^s$, dec.(1950) +64°15'±35', although verification of
the discovery in the second Cambridge Catalogue gave a position
of R.A.(1950) $0^h22^m46^s\pm5^s$, dec.(1950) +63°57'±5'. On the basis of
a further improvement of position in the third Cambridge Catalogue
with a correction applied by Fomalont et al (1974),and recent
high-resolution maps of the source, the centroid of radio emission
is now taken as R.A.(1950) $0^h22^m37^s$, dec.(1950) +63°52'40", and
the source diameter as 7.9 arc min. Although the declination
agrees exactly with Tycho's measurement of the supernova position,
there has been a great deal of speculation and comment on the
error of almost 4 arc min in Tycho's right ascension result (for
a useful discussion see Minkowski, 1968).

No. (Ptolemy)	Identification	Distance
1	ʒ Cas.	10°22'
2	α Cas.	7 50½
3	η Cas.	6 53
4	γ Cas.	5 02
5	δ Cas.	8 03½
6	ε Cas.	9 48
7	ι Cas.	12 58½
11	κ Cas.	1 31
12	β Cas.	5 19

Table 10.3. Tycho's measurements of the angular
distance of the new star from some of the principal stars of Cassiopeia.

Today no one would doubt the identification of the radio source
(3C10:G120.1+1.4) as the remnant of Tycho's supernova, but the
problem of the discrepancy in the position still remains. We have
investigated the situation in detail in the hope of obtaining
a satisfactory solution. Let us start with the observations
themselves.

When the new star appeared, Tycho Brahe had recently finished
making a new sextant graduated in minutes of arc. With this
instrument he proceeded to measure with extreme care over a period
of several months the angular separation of the star from nine
of the principal stars of Cassiopeia. His final results are
given in chapter 4 of the Progymnasmata in a table entitled:
"Distances of the new star from some of the principal fixed stars
in the constellation of Cassiopeia, diligently investigated and
verified". The essentials of this table are duplicated in Table
10.3. Tycho gave the number and name of each star according to
Ptolemy's catalogue. We have identified the stars in modern
notation - by their Greek letter.

Tycho also measured the distance of the new star form the Pole
Star, and he obtained the result 25°14'. Unfortunately virtually
none of his original measurements have survived, although we do
know that he had to make large corrections to his results for
parallax due to the use of a primitive form of sighting. Examples
are quoted by Dreyer (1890): the correction for a measured angle
of 5° was 8' and for an angle of 8° was 13', a roughly linear
relationship. However, we have no way of testing for systematic
errors. In later years Tycho made a series of measurements of the
distances of all the 9 reference stars in Cassiopeia from one
another. His results are given in chapter 4 of the Progymnasmata,
but he tells us that only a small(and unspecified proportion) of
the results were obtained with the original instrument.

It is still of interest to examine the general accuracy of Tycho's
measurements. We have calculated the distances of the 9 reference
stars from one another - a total of 36 results. The mean error in
Tycho's figures was 1.1', a remarkable achievement. Errors of
greater than 2' were rare, although there was one discrepancy as
large as 4.5'.

Our analysis of Tycho's measurements of the distance of the new
star from Polaris and the stars of Cassiopeia is shown in Fig.
10.3. Proper motions have been allowed for, but since the observations were made at various (and unknown) times, no corrections
could be made for differential refraction, etc. Provided the
altitudes were reasonably high the effects of refraction would be
negligible. In the diagram, each line represents the equation

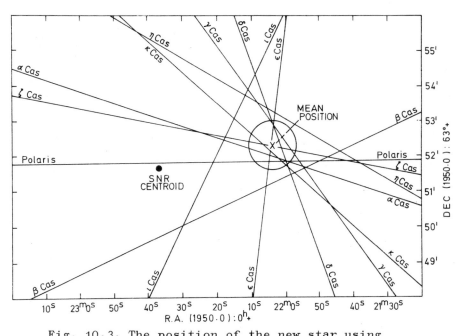

Fig. 10.3. The position of the new star using
Tycho's measurements.

between the R.A. and dec. of the new star deduced from Tycho's measured distances. Over the small angular extent of the figure (only a very few minutes of arc) the various lines may be regarded as accurately straight. Ideally, all the lines would be expected to meet in a single point, which would be the location of the new star. This, of course, is not the case in practice owing to random errors of observation and systematic observational and instrumental errors. The small cross near the centre of the figure represents the best least squares fit to Tycho's data, and the circle round it the corresponding standard deviation in the position. It is obvious that the mutual consistency of the individual results is very good, but the final position (R.A. $0^h22^m3.9^s$; dec.$+63°52'17"$ at epoch 1950.0, with a standard error in both coordinates of $\pm 39"$), although in excellent agreement with the previous determination cited above, is some 4 arc min to the west of the centroid of the supernova remnant (shown by a small shaded circle). There must be some systematic error present since proper motion of the SNR is negligible.

If we examine the distribution of the reference stars in relation to the supernova (see Fig. 10.4), we can see that apart from Polaris, all the stars lie within an angle of less than 180 degrees (actually 140 degrees). It so happens that there are no

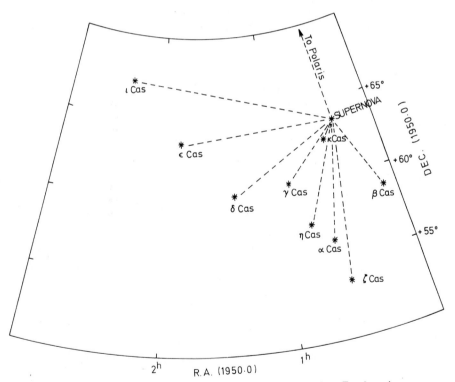

Fig. 10.4. The reference stars used by Tycho in positioning the new star.

nearby stars to the north and west of the position of the supernova, so that any systematic errors would not be expected to
reveal themselves. Is it possible to deduce an improved location
based on the knowledge that there is a significant discord
between the measured position of the new star and the centroid
of the radio source? To test this,we calculated the distance
between this centroid and the nine reference stars of Cassiopeia
(omitting Polaris since the separation is nearly twice any other
value) and compared these figures with Tycho's own measurements.
The results of this investigation are summarised in Table 10.4.

	Distance		
Star	Calculated	Tycho's	Diff.(C-T)
α Cas.	7°49.1'	7°50.5'	-1.4'
β Cas.	5 20.8	5 19.0	+1.8
γ Cas.	4 58.4	5 02.0	-3.6
δ Cas.	7 59.6	8 03.5	-3.9
ε Cas.	9 44.5	9 48.0	-3.5
ʒ Cas.	10 20.7	10 22.0	-1.3
η Cas.	6 49.9	6 53.0	-3.1
ι Cas.	12 57.1	12 58.5	-1.4
κ Cas.	1 28.3	1 31.0	-2.7

Table 10.4. Comparison between the calculated
distance of the SNR from the reference stars
of Cassiopeia with Tycho's measurements.

The mean of all the differences in Table 10.4 is -2.1'. Apart
from   β Cas, all the deviations are within about 1 arc min of
the mean. The result for β Cas is clearly discordant, although
no more so than already noted above for an other measurement.
The mean of all the other deviations, which seem to be independent of the angular separation, is -2.6'. Fig. 10.5 is essentially
similar to Fig. 10.3 above, but a correction of -2.6' is applied
to all Tycho's measurements (with the exception of β Cas and
Polaris, which are discarded). The new mean position is once
more shown as a cross, and the standard error as the circle
surrounding this. The position deduced for the new star is R.A.
$0^h22^m32.3^s$, dec.+63°51'17" with a standard error in both coordinates of +56".
In Fig. 10.5, the SNR lies well within the circle of uncertainty,
so that the single assumption of a systematic error is justified.
This might possibly have arisen through a very slight warping
of the wood of the sextant after construction or through Tycho's
observational procedure in aligning the edges of the sights on
the new star and the comparison star. Whatever the explanation,
it should be remembered just how small an error of 2.6' is
when observing with the unaided eye. Indeed, even the uncorrected
position derived directly from Tycho's data lies within the
boundary of the discrete radio source which is the remnant of
the supernova.

Virtually nothing has been said so far regarding the measurements
of position by Tycho's contemporaries. Many of these results are
very crude being up to half a degree in error or even more. Others
are reasonably accurate, but this may be merely fortuitous.

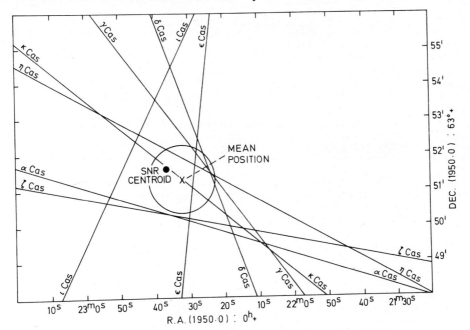

Fig. 10.5. The position of the new star, using
Tycho's measurements with correction for systematic error.

Only the results of three observers deserve special mention.

Mugnoz of Valencia measured the distance of the supernova from
the three stars α, β and γ Cas. His results deviated from the
calculated values for the supernova remnant by 0.9', 0.8' and
11.6' respectively. Clearly he was a fairly careful observer,
although not up to Tycho's standard.

The English mathematician, Thomas Digges of Cambridge, measured
the angular separation of the new star from 6 stars of Cassiopeia.
These were α, β, γ, δ, ε and κ Cas. He used a cross staff for this
purpose. His results are summarised in Table 10.5, with the
calculated values for the SNR for comparison. They are taken
from chapter 9 of Tycho's Progymnasmata.

Digges' results show only a slightly greater scatter than those
of Tycho, and furthermore reveal no evidence of systematic error.
Tycho, comparing his own measurements with those of Digges,
believed that the latter did not make adequate allowance for the
effect of parallax in using his instrument. In point of fact
Digges' results would give a much better location for the supernova than Tycho's. However, this conclusion is only apparent in
retrospect from our knowledge of the accurate position of the
centroid of the remnant. It was natural for Baade and the other
early investigators of the location of the new star to concentrate on Tycho's data in view of his undoubted supremacy as an
observer.

Star	Calculated	Digges	Diff.(C-D)
α Cas.	$7^o49.1'$	$7^o47'$	+2.1'
β Cas.	5 20.8	5 15	+5.8
γ Cas.	4 58.4	4 58	+0.4
δ Cas.	7 59.6	8 05	-5.4
ε Cas.	9 44.5	9 44	-0.5
κ Cas.	1 28.3	1 28½	-0.2

Table 10.5. Comparison between calculated and Thomas Digges' measurements of the new star from Digges' selected stars in Cassiopeia.

Thomas Digges also adopted a very simple procedure which was capable of yielding very accurate results. In order to determine whether the star had any detectable parallax, he selected two pairs of stars so that (as accurately as could be judged) the new star lay at the point of intersection of the lines (i.e great circles) joining each pair. Digges checked the alignment with a ruler 6 feet in length. He stated that the new star lay exactly at the point of intersection of the lines joining δ Cas to β Cep and ι Cep to γ Cas, and concluded that the star could not have a parallax as large as 2 arc min. Tycho, in deducing the position of the new star from Digges' data, showed that Digges had given the stars in the wrong order; the two lines must be understood to join β Cep to α Cas and ι Cep to δ Cas. This correction gave a location very near his own position.

Maestlin adopted a very similar procedure to Digges (Prog., chapter 8), using a taut thread instead of a ruler. He found that the new star lay at the point of intersection of the lines joining δ Cas to ι Cep and β Cas to λ U Maj. His conclusion from this experiment was that the new star had no parallax and was situated among the fixed stars.

We have deduced the 1950.0 coordinates of the supernova by combining the observations of Digges and Maestlin (it will be noticed that both observers chose the pair of stars δ Cas and ι Cep). All three lines meet almost in a single point, having coordinates approximately $0^h23^m13^s$, $+63^o50.6'$. However, this is nearly 4 arc min to the east of the centroid of the SNR. In view of the difficulty in determining whether the ruler or thread was absolutely straight, this seems a remarkably good result. The discrepancy is in the opposite direction to that of Tycho. To sum up, there seems more than adequate grounds for believing that the original measurements of the position of the supernova are in full accord with the modern determination of the location of the SNR centroid.

Figures 10.6 and 10.7 show high-resolution maps of the radio remnant of Tycho's supernova made with the Westerbork Synthesis Radiotelescope at frequencies of 610 and 4995 MHz respectively (Duin and Strom, 1975). At the lower frequency the resolution of the instrument is ≈1 arc min, and at the higher frequency ≈7 arc sec. The lower frequency map shows a well-defined ring structure characteristic of supernova remnants, although in the

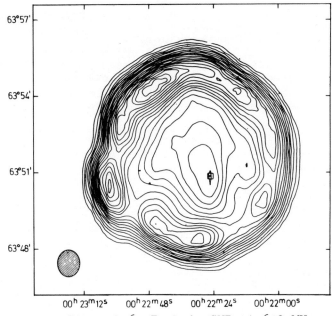

Fig. 10.6. Tycho's SNR at 610 MHz.

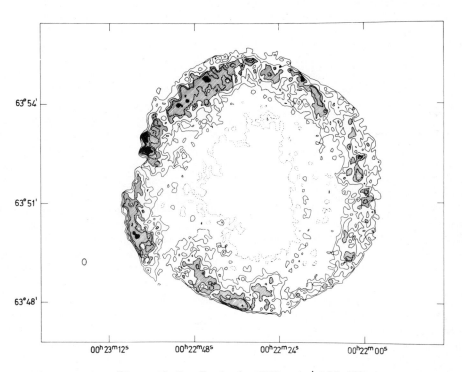

Fig. 10.7. Tycho's SNR at 4995 MHz.

high-resolution map the ring breaks up into emission "knots" -
this structure is consistent with the radio emission mechanism
of Gull (1973a) for young supernova remnants, where convective
mixing at the interface of the ejecta and swept-up interstellar
material produces local enhancements of the magnetic field and
consequently radio brightness. (The central minimum of Fig. 10.7
is exaggerated because of analysis difficulties.)

The radio spectral index of the remnant has, until recently,
been interpreted as steepening towards low frequencies, with
α =-0.55 at $\approx$1000 MHz and α =-0.9 at frequencies <100 MHz.
Roger et al (1973) suggested that the data could be fitted with
two power law spectra having equal flux densities at 500 MHz,
and spectral indices of -0.95 and -0.35. The new high resolution
data of Duin and Strom in fact suggest that these components
are spatially separated. Their spectral index varies across the
source, being $\approx$-0.7 for features associated with the outer shell,
steppening to -1.0 as one proceeds radially inwards, suggesting
that the high energy electrons producing the synchrotron emission
are distributed in a thinner shell than the low energy ones.

Polarisation measurements at different frequencies have in the
past appeared inconsistent, mainly due to the unknown effects
of Faraday rotation. Duin and Strom were able to determine
"zones" of rotation measure from a comparison of polarisation
observations made at 1420 and 5000 MHz, and correcting for this
rotation their map of intrinsic polarization is shown as Fig.
10.4. The tangential alignment of the electric field vector
around the entire shell, from the outer to innermost radii, is
seen to be extremely good confirming that, as in other young
SNRs (e.g the remnant of AD 1006, see chapter 7), the projected
magnetic field perpendicular to the line of sight is directed
almost perfectly radially everywhere,

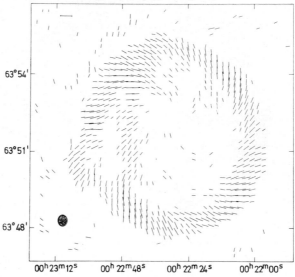

Fig. 10.8. Tycho's SNR, showing intrinsic polarization.

The distance to the remnant is stll a cause of some debate. Menon
and Williams (1966) suggested a distance of ≈5 kpc from H I
absorption data, giving a linear diameter of ≈11 pc and average
velocity of expansion since the outburst of ≈13,400 km/sec.
Comparable values had earlier been suggested by Minkowski (1964)
and Woltjer (1964) from estimates of maximum brightness of the
supernova. However Hughes et al (1971) suggested that the hydrogen distribution in the vicinity of the remnant was anomalous,
and interpreted their H I absorption data as merely requiring a
minimum distance of 1.5 kpc. More recent observations do in fact
confirm earlier estimates of somewhat greater required minimum
distance. Williams (1973) gives a distance ~5 kpc, while Goss
et al (1973) interpret a weak absorption feature not apparent in
some earlier observations because of observational limitations
as requiring a distance ⩾6 kpc. In interpreting the dynamical
evolution of supernova remnants in Chapter 12 we have cause to
adopt this larger value, so that the linear diameter is ~13.8
pc and the average velocity of expansion since the outburst
~16,000 km/sec - an extremely large value, even for a Type I
supernova.

Referring to Fig. 1.5, estimates of an apparent magnitude at
maximum of -4 require an average interstellar absorption along
the line of sight of a mere 0.2 mag./kpc assuming a typical
Type I peak absolute magnitude of -19. While this value is somewhat lower than expected for this direction in the galactic plane
it does not represent a serious discrepancy as to call into
question the H I absorption measurement distance. Supporting
evidence for such a low absorption is given by van den Bergh
(1970), who estimated the reddening of the supernova from the
colour descriptions in the historical records; from a comparison
of the observed colours with the intrinsic colours of Type I
supernovae he determined an interstellar absorption of about 1.6
magnitudes. For a distance of ⩾6 kpc, this would require a peak
absolute magnitude of -19.5 or brighter. We will present later
evidence suggesting that Tycho's supernova was a particularly
energetic outburst, so that the peak absolute magnitude may
indeed have been exceptionally large.

The optical remnant defied detection until an accurate position
was determined from radio observations. Then Minkowski (1959)
found extremely faint filaments seen in Hα and concentrated
along the outer rim of the radio shell; these are usually interpreted as emitting sheets seen edge-on (van den Bergh et al,
1973). Only one filament is bright enough to permit spectroscopic
observations. Searches for prominent emission lines such as O II
and O III have been unsuccessful, and since interstellar absorption in the optical is low (see above) any such emission must
therefore be intrinsically weak. Comparison of plates taken over
the past two decades show that individual filaments display
significant changes in brightness, and are moving radially at a
rate of ≈0.2" per year. This is somewhat slower than the required
average motion since the outburst of ≈0.6" per year, however it
now appears that optical filaments in remnants are shocked interstellar clouds which may not directly exhibit the properties of
the expanding shock wave (McKee and Cowie, 1975). In the absence

of observable details near the centre of the remnant, the radial
velocity of expansion cannot be determined so that it is not
possible to reach any conclusions about the distance to the
remnant from optical measurements.

A relatively bright star located near the centroid of the radio
emission shows a late-type spectrum and is clearly not the
stellar remnant. Van den Bergh et al (1973) report that attempts
to obtain a spectrum of a faint bluish companion to this star
have so far remained tantalizingly unsuccessful.

The remnant of Tycho's supernova is a weak X-ray source first
detected by Friedman et al (1967). It is found from proportional
counter data that the spectrum of the emission is well fitted
by a two-component thermal model (Davison et al, 1976), with a
prominent emission feature attributed to iron. Thermal emission
from a shock-heated interstellar material must therefore now be
accepted as the source of the X-ray emission (contrary to earlier
suggestions by Shklovsky, 1974, who suggested that the X-ray
source was most likely a pulsar). The emission feature requires
an abundance of iron in the neighbourhood of the remnant 10 to 20
times greater than the typical interstellar value; this could
be achieved by enrichment of the iron content by material ejected
from the supernova having an iron abundance four times the cosmic
value (Davison et al, 1976). Higher resolution spatial and
spectral observation of the X-ray source are urgently required.

The historical observation reported by Tycho Brahe provide us
with more primary data for the AD 1572 supernova than that combined from all previously recorded outbursts. In particular the
unambiguous classification of the event as of Type I, and the
positive identification of its remnant at optical, radio, and
X-ray wavelengths, provide astrophysicists today with a near-
ideal candidate for studying the behaviour of a "typical"
(although particularly energetic) SNR. It is to be expected
that observational advances will in the near future resolve any
remaining uncertainties as to the distance of the remnant, the
identification of the stellar remnant, the nature of the optical
filamentary emission, and the structure of the X-ray source.

In AD 1596, after falling out with his patrons in the Danish
Royal Court, Tycho Brahe left his native Denmark to become
Imperial Mathematician of Bohemia. He was soon joined by a much
younger man by the name of Johannes Kepler, who had been born
just 11 months prior to the recorded outburst of the supernova
which was to make Tycho famous. It must rank as one of astronomy's
most fortunate coincidences that the next galactic supernova was
to be sighted during Kepler's lifetime, but one of astronomy's
greatest tragedies that its appearance should precede the applic-
ation of the telescope to astronomical research by a mere five
years.

Chapter 11

THE NEW STAR OF JOHANNES KEPLER

Many people who witnessed the AD 1572 event must have also seen
this star, which occurred only 32 years later. Unfortunately,
Tycho Brahe, who did so much for the earlier new star, had died
in AD 1601. However, his successor as Imperial Mathematician at
Prague, Johannes Kepler (AD 1571 - 1630), was to make similar
detailed observations of the new star which bears his name.
These enable us to prove that the star, like its predecessor,
was a supernova of Type I, and conclusively identify its remnant
as G4.5+6.8, an established SNR. Kepler's greatest work was
undoubtedly his study of Tycho's observations of the planet
Mars, which led to the formulation of his laws of planetary
motion. However, his study of the supernova must also rank as
one of the major contributions to science of the age.

Despite its southerly declination (-21°) and proximity to the Sun
(only 58° E at discovery), the new star was detected almost
simultaneously by astronomers in Europe and the Far East.
It was a fortunate coincidence that the star appeared only about
3 degrees to the northwest of Mars and Jupiter (which were then in
conjunction) and about 4 degrees to the east of Saturn, so that
this area of sky would be under constant surveillance at this
time. The supernova was discovered on the evening of October 9
(all dates are on the Gregorian calendar) independently by two
Italian observers. Baade (1943) points out that "we have the
emphatic statements of Fabricius and several other reliable
observers that nothing extraordinary was noticed when they
observed Mars and Jupiter on the evening of October 8". The star
was thus still inconspicuous on October 8. By October 10, the
star was sighted in China, and three days later the court
astronomers of Korea (at Sŏul) first saw it. Bad weather in
Northern Europe prevented observation on several successive nights.
For example, although Kepler had been notified of the discovery of
the supernova on October 10, he did not himself see it until a
full week later. Possibly the Korean astronomers, who, as we shall
learn, so carefully observed this star, were similarly troubled
for the first few days.

As will be shown below, the probable date of the maximum of the
supernova was around November 1, so that the star was discovered
nearly 3 weeks before the maximum. Even for the systematic supernova searches maintained during the past few decades at Mt. Palomar,
Asiago, etc., such a large interval is quite rare.

Conjunction of the star with the Sun occurred on December 9,
but after heliacal rising, it was recovered both in Europe and
in the Far East. The last definite sighting by Kepler was on
October 8 in the following year, only two days after the star

was last seen in China.   The period of visibility was thus virtually a full year.

The position and changing brightness of the new star were measured very carefully by the European and Korean astronomers. It seems possible that similar measurements were made in China, but only brief records have survived. As is true of the AD 1572 event, there is no known reference to the star in Japanese history.

Following our usual practice, we shall begin with the Far Eastern records. Chinese observations of the new star are reported in the astronomical treatise of the Ming-shih (chapter 27) and in the Ming-shih-lu (Annals of the reign of Emperor Shên-tsung). Ming-shih: "32nd year (of the Wan-li reign period of Emperor Shên-tsung), 9th month, (day) i-ch'ou. In the division (fên) of Wei there was a star like a crossbow pellet. Its colour was orange (i.e. reddish-yellow) and it was seen in the south-west. In the 10th month it became obscured. On (the day) hsin-yu in the 12th month it reappeared in the south-east. It was still in the division of Wei. In the 2nd month of the following year it gradually became dim. On (the day) ting-mao in the 8th month it was finally extinguished".
Ming-shih-lu: "(32nd year of the Wan-li reign period, 9th month), (day) i-ch'ou. At the beginning of the night in the south-west there arose a strange star as large as a crossbow pellet. Its body was orange (i.e. reddish-yellow) in colour. It was called a 'guest star' ".
Ming-shih-lu: "(33rd year of the Wan-li reign period, 8th month), (day) ting-mao. At night the guest star was not seen. From the 9th month of the 32nd year the guest star was visible in the division of Wei. At the first watch it regularly appeared in the south-west, and it turned with the sky. During the 10th month in the evening it sank and was not visible. In the 11th month at the 5th watch it regularly appeared in the south-east. During the 1st month of the present year its light gradually became dim and now it has finally been extinguished."

Combining these three records, the date of first appearance of the star corresponds to AD 1604 October 10. The date when the new star was "finally extinguished" corresponds to AD 1605 October 7. Heliacal setting took place in the month between November 21 and December 21 (in AD 1604). There is discord between the Ming-shih and the Ming-shih-lu over the date of heliacal rising and the month when the star "gradually became dim". The latter is of course unimportant since the interpretation of "dim" (an) is quite subjective. The Ming-shih gives the exact date of heliacal rising (equivalent to AD 1605 February 3), but this seems very late. The Ming-shih-lu date (December 22 to January 18) agrees well with the Korean date (December 22 - see below). There is an unaccountable error somewhere.

The Chinese records would alone be sufficient to prove that the star was not a comet, but only a very vague location is given, i.e. an approximate R.A., but no dec. A rough dec. might be deduced from the direction in which the star was seen, but it seems unlikely that the remnant could be located from the Chinese record

alone.   There is even no direct reference to the brightness of the
star.

The standard of all three Chinese records is uniformly low.
We find it difficult to judge whether the original observations
of the imperial astronomers (at Peking) were deficient or whether
the chronicles omit much valuable material in the interests of
brevity.

In contrast, the regularity and detail of the Korean observations
come as a major surprise.   Only a very brief Korean record of
SN 1572 exists, but for nearly 6 months we find almost daily
reports of the new star of AD 1604.   These are contained in the
Sŏnjo Sillok, the annals of the reign of King Sŏnjo of Korea
(AD 1567 to 1607).   On reading the reports of the star in the
Sŏnjo Sillok, one cannot fail to be impressed by the remarkably
scientific attitude of the court astronomers.   On almost every
clear night the right ascension of the star and its north polar
distance were measured, while its brightness was compared with that
of neighbouring planets and stars, just as in Europe.   We are
regularly told when thick clouds blanketed the sky, ruling out
observation.   Frequent statements to the effect that cloud or
bright moonlight, although not conclealing the star, prevented
measurement of its co-ordinates, testify to the care which was
exercised.   Although positions are only determined to the nearest
degree, after the first night the same location is reported night
after night.   Almost one hundred observations of the new star are
recorded, and we are told that cloud either wholly or partially
interfered with visibility on nearly 40 nights.   This day to day
diary of observations of the star is without equal anywhere else
in the world.

Like most Korean literary works, the Sŏnjo Sillok is written in
Classical Chinese (Hanmun).   Because of this, we have preferred to
give the romanisation of characters as for a Chinese work except
for the name of the king (Sŏnjo), for which the Korean pronun-
ciation is given.

The first report of the new star is as follows: "(37th year of
Sŏnjo, 9th month), (day) wu-ch'en.   In the first watch of the might
a guest star was in the 10th degree (tu) of Wei lunar mansion and
distant 110 degrees from the (north) pole.   Its form was slightly
smaller than Jupiter.   Its colour was orange (i.e. yellowish-red)
and it was scintillating.   In the 5th watch there was mist".

The discovery date corresponds to AD 1604 October 13.   On the
following night a slightly revised position is given ("the 11th
degree of Wei and distant 109 degrees from the pole), and the detail is added that the guest star was "above T'ien-chiang".   Otherwise the record is virtually identical with the previous one.
On all future occasions this revised position is retained, and
apart from changes in the brightness of the star, essentially this
same account is found on every clear night up to November 26.
Between October 28 and November 5, the star was compared with
Venus in brightness, and from October 28 to 31 we find the description "its ray emanations were very resplendent", suggesting

that it was at maximum brilliance around this time.

After November 26 there were a few cloudy days - "there were
dense clouds and the guest star was not seen". There are no
mentions of the star from November 30 to December 3, and by December
4 the star had already set heliacally. On this date the entry
reads:"The guest star was close to the Sun. It sank in the west
before dusk and could not be observed". The precise date of
heliacal setting is thus unknown, but it was sometime between
November 27 and December 3.

By December 27 the star was again visible. For this day we read,
"At daybreak the guest star was seen in the east above the stars
of T'ien-chiang. It was within the 11th degree of Wei and distant
109 degrees from the pole. Its form was larger than Antares.
Its colour was orange and it was scintillating". Apart from
slight textual variations and the diminishing brightness of the
star, precisely the same account occurs on every clear night
except when the Moon was bright until April 5. After this date,
references to the star are few and far between, presumably because
the original observations had been lost. The single record on
April 23 follows the usual pattern, while on May 2 we read,
"at the 4th watch the guest star was faintly visible through
rifts in the clouds". Subsequently, there are only three indirect
references to the star, two commenting on its astrological significance, and the last, dated September 15, stating that "there
were dense clouds and the guest star was not seen". It thus seems
probable that the star continued to be observed for a similar length
of time to that in China and Europe (i.e. for virtually a full
year).

The Korean measurements of position are interesting in that they
prove that the position of the star remained fixed for nearly
6 months. However, their accuracy is far inferior (by two orders
of magnitude) to the European measurements. Nevertheless, we have
thought it of interest to compare the Korean location with that
derived from the European measurements (see below). From chapter
2, the determinant star of Wei is μ'Sco. The R.A. of the new star
thus reduces to 17^h 06.6^m $+ \overline{2^m}$. This compares with the best
European result of 17^h $07.\overline{1}^m$. The Korean position is thus tolerably
accurate. The polar distance (109 tu) corresponds to a declinat-
ion of -17.4° (European result $-21.\overline{1}^\circ$) so that the error is thus
rather large (some 4 degrees). If this distance was measured
from Polaris, which would presumably be the pole star of the time,
the error would be slightly larger, so that this is not an explanation.

A remarkable observation made by the Korean astronomers on January
20 in AD 1605 allows us to fix the position of the star with a
fair degree of accuracy from the oriental data alone. The
appropriate entry in the Sŏnjo Sillok reads : "(37th year of
Sŏnjo, 12th month), (day) ting-wei ...... Venus invaded the guest
star (chin-hsing-fan-k'o-hsing)". As discussed in Chapter 9, the
expression fan ("to invade", "trespass against", etc.) was an
astrological term implying an approach of two celestial bodies to
within about a degree of one another. Calculation based on the

European position shows that before dawn on January 21 (astronomical date January 20 - see Chapter 9), Venus and the new star would be only 0.51 degrees apart. In the absence of the European measurements this single observation would probably have been sufficiently accurate to allow the remnant of the new star to be identified with confidence. Fortunately, thanks to the efforts of Kepler and Fabricius, this is no more than an academic exercise.

By far the most important Korean observations are the brightness comparisons. Before heliacal setting the brightness was compared with the planets: Jupiter (sui-hsing - "Year Star"), Venus (chin-hsing - "Metal Star") and Mars (hui-hsing - "Fire Star"). After heliacal rising, only two reference stars were chosen : (α Sco. (Hsin-ta-hsing - the "Great Star of Hsin") and τ Sco. (Hsin-tung-hsing - the "Eastern Star of Hsin"). The lunar mansion Hsin consists only of three stars, with τ Sco at the east, α Sco (Antares) in the middle and σ Sco at the west so that the Korean names are unambiguous. The new star was described as "larger than" (ta-yu) Antares between December 26 and January 20, "as large as" (ta-ji) Antares between January 21 and February 15, "slightly smaller than" (ch'a-hsiao-yu) Antares between February 19 and 21, and finally "smaller than" (hsiao-yu) Antares between February 23 and April 5. Eventually, on April 23 the star was described as "as large as" (ta-ju) τ Sco.

We have corrected two minor presumed errors relating to January 30 and February 18. On both these dates the star was described as "larger than" (ta-yu) Antares, which does not conform with the general trend. Possibly these are textual errors. Alternatively, local atmospheric conditions could have made the new star seem rather brighter than Antares on each occasion. It seems best to regard both reports as spurious.

It so happens that Antares was an unfortunate choice for it is a semi-regular variable. The approximate period is 5 years and the extreme range of magnitude, reduced to the Harvard system, +1.1 to 2.0 (Kukarkin et al, 1969). In studying the light curve of the new star we have assumed this same range of brightness.

The various brightness comparisons are summarised in Table 11.1.

In Table 11.1 dates do not always run continuously, either because of gaps in the observations (due usually to cloud) or in the records themselves (presumably due to loss of the original reports). Although the new star was compared with Venus during the period October 28 to November 5, at that time the planet was a morning star (on the opposite side of the Sun). Direct comparison, unless in daylight- and there is no reference to the visibility of the new star during the daytime - was thus impossible, and the only sound inference seems to be that the star was considerably brighter than Jupiter. This is how we shall interpret the observations.

Mars was the reference object in the last few days before heliacal setting (November 20 - 26). However, the planet was then about 50 degrees east of the Sun while the star was rapidly approaching

DATE OR DATE RANGE	BRIGHTNESS LEVEL
AD 1604 Oct. 13/14	Smaller than Jupiter
Oct. 15/16	Slightly smaller than Jupiter
Oct. 17/18	As large as Jupiter
Oct. 19-25	Larger than Jupiter
Oct. 28-31	As large as Venus, rays prominent
Nov. 1- 5	As large as Venus
Nov. 7-13	As large as Jupiter
Nov. 14	Slightly smaller than Jupiter
Nov. 16-19	Smaller than Jupiter
Nov. 20-23	As large as Mars
Nov. 24-26	Smaller than Mars
Dec. 26-Jan. 20	Larger than Antares
AD 1605 Jan. 21-Feb. 15	As large as Antares
Feb. 19-21	Slightly smaller than Antares
Feb. 23-Apr. 5	Smaller than Antares
Apr. 23	As large as τ Sco

Table 11.1. Comparisons of the brightness of the new star
with planets and fixed stars by the Korean astronomers.

conjunction (elongation only 13 degrees east on November 26).
The relative visibility of the new star would thus be very much
impaired by atmospheric extinction (on account of its low altitude)
and the twilight glow, giving the impression that it was much
fainter than it actually was. When the comparisons with Jupiter
were made, planet and star were fairly close together so that
both would be viewed under much the same conditions.

Before leaving the oriental records, it seems appropriate to make
some reference to the "night watches" (kêng) which were in use
throughout the Far East. Generally speaking, the night between
dusk and dawn was subdivided into five watches, presumably of
equal length. From Needham, Wang-ling and Price (1960), the
durations of dusk and dawn were each assumed to be $2\frac{1}{2}$ k'o (36
minutes) - similar to our "lighting up time". Normally the remainder of the interval between sunset and sunrise would be taken
up by the watches. The average length of a watch would thus be
about 2 hours 10 minutes, but the actual length would obviously
vary with the season. In this respect, the night watches were
similar to "seasonal hours" which were used throughout the
ancient and medieval West.

Coming now to the European observations of the new star, Kepler
made a detailed study of the object in his De Stella Nova in
Pede Serpentarii. However, whereas Tycho Brahe collected practically every significant observation of the supernova of AD 1572
made by his contemporaries, the literature relating to the star
of AD 1604 is very scattered. The best reference work is undoubtedly the paper by Baade (1943). Baade searched carefully through
early 17th century European astronomical writings for observations.
As a result of this search, he felt that he had probably isolated
all important documents with the exception of the reports of
David Fabricius, which were not accessible to him. However,
Baade was able to obtain a few of the observations made by
Fabricius from the latter's correspondence with Kepler so that
this omission was probably not too serious.

It seems that the new star first appeared on the evening of
October 9. On that day it was reported independently by two
Italian observers - a physician in Cosenza whose name is unknown
and I. Altobelli in Verona. The physician reported his observations
of the star to the Jesuit astronomer Clavius in Rome. As already
noted, we have the testimony of Fabricius and others, who observed
Mars and Jupiter on the evening of October 8, that the star was
not then evident. By October 10, the new star was seen by two
observers in Padua, and also in Prague by J. Brunowski, who
notified Kepler of his discovery. Baade collected the various
estimates together in a table. This is reproduced as Table 11.2,
but we have excluded the names of the observers and the few references to the colour of the star. On October 9 the physician
reported the colour of the star as "like Mars", and on the same day
Altobelli made the analogy "like half of a ripe orange". Again
Capra-Marius of Padua on October 10 made a comparison with Mars,
but by October 15, both the physician and Fabricius described
the star as "like Jupiter", Fabricius adding that it was "white,
not red". These latter colour estimates are at variance with

those made in Korea, which were uniformly "orange" from discovery
until AD 1605 April 5.   On April 23 the colour was said by the
Korean astronomers to be "yellow", but possibly by this time the
star would be so faint that colour judgement would be unreliable.

Date	Brightness estimate
AD 1604 Oct. 8	Not seen
Oct. 9	As bright as Mars
"	As bright as Jupiter
Oct. 10	Somewhat brighter than Mars
"	Very similar in brightness to Mars
Oct. 11	Still brighter than on Oct. 10 when somewhat brighter than Mars.
"	Twice as bright as Jupiter
Oct. 12	Almost as bright as Jupiter
Oct. 15	As bright as Jupiter or somewhat more
"	Much brighter than Jupiter; no further increase after this day
"	As bright as Jupiter or a little more; no further increase afterward
"	A little brighter than Jupiter
"	As bright as or brighter than Jupiter
"	Brighter than Jupiter and equal to Venus
Oct. 17	Much brighter than Jupiter (almost twice as bright)
AD 1605 Jan. 3	Brighter than α Sco , much fainter than α Boo
Jan. 13	Brighter than α Boo and Saturn
Jan. 14	About as bright as Mars (in Oct. 1604)
Jan. 21	About as bright as α Sco , a little brighter than Saturn
End of Jan.	As bright as α Vir
Mar. 20	Not much brighter than $_3$ and η Oph
Mar. 27	Not much brighter than $_3$ and η Oph
Mar. 28	Not much brighter than η Oph
Apr. 12	As bright as η Oph
Apr. 21	As bright as η Oph
Aug. 12–14	As bright as $_3$ Oph
Aug. 29	About as bright as $_3$ Oph
Sep. 13	Fainter than $_3$ Oph
Oct. 8	Difficult to see; fainter or equal to $_3$ Oph

Table 11.2. European estimates of the brightness of the new star.

Most of the later observations in Table 11.2 (after the beginning
of AD 1605) were made by Kepler.   There do not appear to be any
useful European estimates of the brightness of the star from
October 17 until January 3.   Kepler (De Stella Nova, chapter 1)
judged that the star "was seen with almost the same magnitude
during the whole of the month of October".   Before conjunction
with the Sun he last saw the star on the evening of November 16.

He was frustrated by cloud on the next few days, and by November
22, although, as he tells us, the position of the star was indicated by the presence of the Moon (then a young crescent), he could
not see it. Kepler concluded that from Prague (50° N) the new
star set heliacally between November 16 and 22. In Korea, the
date of heliacal setting was some time between November 27 and
December 3, but the oriental astronomers had the benefit of a
much more southerly location (37½° N).

It was not until January 3 (AD 1605) that Kepler rediscovered the
star (after conjunction with the Sun). This was a full week after
the Korean astronomers sighted it, but apart from his inferior
situation, Kepler had been clouded out for some time. He noted
that "it was much lessened from its original magnitude". After
this we only get occasional bulletins regarding the brightness
of the new star, usually at intervals of about a month, but there
is a major gap between April 11 and August 12 during which apparently he made no observations. Kepler does not appear to have
observed the star anything like so regularly as Tycho Brahe some
years before.

The last positive sighting was on October 8 in AD 1605. Kepler
wrote: "Now exactly a year after its first apparition in a very
clear sky, its appearance could be noted only with difficulty ...
From this time anew it set heliacally".

At the beginning of AD 1606 the inclement weather gave Kepler
few opportunities to search for the star. He looked carefully
for it on the morning of January 26, and again on February 4
but could not detect it. His comment on February 6 is interesting:
"I left the observatory, not sure whether I had seen any trace of
the new star. Therefore it seems to have become too small to be
seen even on this clear morning, if it has survived". After a
further search in March, having allowed time for the elongation of
the star from the Sun to increase, he realised that it had ceased
to be visible. His final conclusion was that "It is therefore
uncertain on which day between October in the year 1605 and
February in the year 1606 it had vanished".

In both the European and Korean brightness estimates we find
expressions equivalent to some of the following: "much brighter
than"; "brighter than"; "slightly brighter than"; "as bright
as"; "slightly fainter than"; "fainter than". In making his
magnitude reductions, Baade adopted a photometric step of 0.25
mag. between each of the above intervals. Thus we have the following conversions:

Much brighter than	-0.75 mag.
Brighter than	-0.5
Slightly brighter than	-0.25
As bright as	0
Slightly fainter than	+0.25
Fainter than	+0.5

Baade justified his choice of magnitude interval in two ways.
Modern variable star observers adopt a photometric step of 0.1

mag., but such a high accuracy would be unlikely to be achieved
by relatively untrained observers. A better estimate is provided
by the smallest magnitude interval used in European star catalogues
around AD 1600. This is one-third of a magnitude, which Baade
considered more appropriate. In practice he reduced this to a
quarter of a magnitude since the supernova was mostly compared
with nearby stars. For the time near maximum, Baade made an attempt to allow for the rapid increase in the width of the photometric step at higher brightness; he multiplies all magnitude
intervals by 1.5. To provide an independent check by an experienced
variable star observer, we asked Dr. Patrick Moore of Selsey
Observatory, England, to comment on the Korean data. He suggested
magnitude conversions which were in excellent agreement with Baade's
choice. We have accepted Baade's figures without further question.

Two factors which Baade somewhat surprisingly ignored in his
magnitude reductions are the effect of moonlight and differential
atmospheric extinction. A close approach of the Moon (particularly
when full) to either the supernova (which was very near the ecliptic)
or the comparison object could appreciably alter the relative
visibility. However, with the aid of the tables of Tuckerman
(1964), we have verified that on the dates of all the available
European observations the Moon did not come within 30 deg. of the
supernova or reference planet/star so that its effect can be
ignored. The Korean observations were, of course, made almost
daily. However, because of this, the only important records are
those for which changes in brightness are noted in the middle of
a period when the star and compared object were judged of equal
brightness. On these critical dates, interference by moonlight
was again negligible.

Differential atmospheric absorption between the supernova and
comparison object on a number of occasions proves to be significant. This largely results from the low meridian altitude of the
star in both Europe and Korea. We have examined each individual
observation to allow for differential atmospheric absorption as
far as possible. In this we have made use of the tables of
Bemporad (1904), as published by Schoenberg (1929). Bemporad's
tabular values have been amended slightly by us to correspond
with atmospheric transmission coefficient of 0.80 quoted by
Allen (1973).

Visual magnitudes of the various comparison stars are on the
Harvard system (cf. Schlesinger, 1940). Obviously, as the object
is to produce a light curve, the precise magnitude system is
relatively unimportant. We have checked on the variability of
each star using the catalogue of Kukarkin et al (1969). Apart
from α Sco, already discussed, all reference stars can be regarded as of constant brightness. Planetary magnitudes -also on the
Harvard System - were computed by Baade and we have accepted these
without alteration. The calculated visual magnitudes of the new star
are listed in Tables 11.3 and 11.4, treating Korean and European
data separately. In each case the comparisons with Venus are understood to mean "much brighter than Jupiter". Observations not included in these tables are the Korean comparisons with Mars and the

European ones with Jupiter on October 9 and 10, α Boo on January 3
and Mars on January 14.   The Korean observations have already been
discussed. Baade justifiably rejected the Jupiter comparisons on
October 9 and 10 since these differ from those of the other observers by more than 2 mag.   Again, the α Boo comparison differs
by more than 2 mag. from neighbouring estimates (allowing for
atmospheric extinction).   Finally the reference to Mars on January 14 must be rejected since it refers to the brightness of the
planet as it was 3 months previously (in October of AD 1604).

In the tables, days are counted after AD 1604 October 8 and magnitudes are estimated to the nearest 0.05.

Day No.	Corrected mag.
6	− 1.1
8	− 1.45
10	− 1.8
11	− 2.55
20	− 2.95
28	− 2.95
33	− 1.95
37	− 1.7
39	− 1.35
104	+ 0.8 (mean)
119	+ 1.55 (mean)
134	+ 1.95 (mean)
138	+ 2.3 (mean)
198	+ 2.9

Table 11.3.   Reduction of Korean brightness
estimates to magnitudes.

Day No.	Corrected mag.
0	+ 3 or fainter
1	+ 0.9
2	+ 0.5
3	− 0.7
4	− 1.5
7	− 2.2
9	− 2.6
87	+ 1.05 (mean)
97	0.0
105	+ 1.55 (mean); + 0.5
114	+ 1.2
164	+ 2.25
171	+ 2.25
172	+ 2.25
187	+ 2.40
196	+ 2.45
310	+ 4.45
326	+ 4.45
341	+ 4.95
366	+ 4.7

Table 11.4.   Reduction of European brightness
estimates to magnitudes.

Fig. 11.1   The light curve of the supernova of
AD 1604 constructed from European (shaded circles)
and Korean (crosses) observations.

The various estimates are shown diagrammatically in Fig. 11.1.
Korean observations are denoted by crosses, European observations
by shaded circles.   In the case of the Antares comparisons,
the range of magnitudes corresponding to the probable extremes
in brightness of the variable are indicated.   The light curve
obtained from such a varied assortment of crude estimates is
remarkably well defined.   This is clearly characteristic of a
Type I supernova, as for the new star of AD 1572.

A fair estimate of the apparent magnitude of the supernova at
maximum is around -3.0.   The date of maximum was probably some 20
days after October 8, and thus about October 28 (in 1604).

From the form of the light curve, the supernova nature of the new
star is well established.   There remains the question of the
remnant.   The angular distance of the supernova from the stars
  α Aql (Altair), α Oph, η Oph, ξ Oph, α Sco (Antares) and σ Sgr
was carefully measured by Kepler.   Fabricius made similar measurements for all of the above stars except σ Sgr.   From this data the
position of the new star was deduced by Schlier (1934) and Böhme
(1937).   The former used only the measures of Fabricius, having
demonstrated, somewhat surprisingly, that these were far superior
to Kepler's data.   His result, reduced to the epoch 1950.0 is:
R.A. $17^h27^m39.1^s \pm 3.1^s$; dec. $-21°26'39'' \pm 42''$.   Böhme, after he
corrected some "obvious misprints" in Kepler's results, solved
from both sets of data.   He obtained a position almost identical
with that derived by Schlier: $17^h27^m38.5^s \pm 2.0^s$; $- 21°26'38''$
$\pm 26''$ at epoch 1950.0.

Using the above positions, Baade (1943) made a search for an
optical remnant.   He used red light because the supernova
appeared in a heavily obscured part of the Milky Way.   He was
immediately successful;   the very first plate revealed a faint
patch of nebulosity whose centre lay only 2.1^s W and 1" N of the
position of the new star.   The positional agreement is thus far
superior to that in the case of Tycho's supernova.

Listed as number 358 in the third Cambridge catalogue, the
remnant of Kepler's supernova is a strong non-thermal radio source
of ~3 arc min diameter.   Its detailed structure remained essentially uncertain until recently, since the southerly declination
puts it beyond the range of objects that can be properly mapped
by high-resolution aperture synthesis radiotelescopes in the
Northern Hemisphere;   (only now are Southern Hemisphere synthesis
radiotelescopes being built with sufficient resolution to study
sources as small as Kepler's SNR).   The beam-shape of an attempted
high-resolution synthesis map by Hermann and Dickel (1973)
was so distorted as to make the map almost impossible to interpret.   A lunar occultation measurement by Hazard and Sutton
(1971) was reported to be consistent with the source having a
broken shell of diameter 3 arc min.   The true structure of the
source was eventually revealed from 5000 MHz observations made
using the Cambridge one-mile radiotelescope and the Owen's
Valley interferometer and analysed by Gull (1975).   Although
each set of observations independently suffered from limited
coverage of the aperture plane, the data could be very

Fig. 11.2   A 5000 MHz map of the radio remnant
of Kepler's supernova, constructed by Gull (1975)
from observations made with the Cambridge one-
mile radiotelescope and the Owen's Valley interferometer.

carefully combined and analysed to produce the map depicted as
Fig. 11.2.

Gull's map of the remnant shows the shell structure characteristic
of SNRs (and hinted at by the earlier lunar occultation observations). Although almost circular, the shell shows peak brightness
along the northern edge, with a large gap to the south and irregular indentations to the east (possibly due to encounters with
irregularities in the interstellar medium). The radio properties
again appear to be consistent with the preferred generation
mechanism of Gull (1973a) for young SNRs.

The centroid of the radio emission, R.A. (1950) $17^h 27^m 41^s$,
Dec. (1950) $-21^o 27'$, is in spectacular agreement (differing
by less than 30 sec of arc) with the positional determination of
Fabricius given above.

The radio spectrum of the remnant is well determined. Milne
(1969) has summarised data from observations in the range 86
to 5000 MHz which indicate a non-thermal spectral index of
$\alpha = -0.58$. However there is no evidence for strong polarisation
in Kepler's SNR (Milne and Dickel, 1974).

There are no observations which enable an unambiguous estimate
of distance to the source. It is too close to the Galactic
centre for H I absorption measurements to produce a meaningful
determination, and the optical remnant (discussed below) is too
diffuse to obtain transverse motions with sufficient precision
to obtain the distance by combining them with radial velocity
estimates (Minkowski, 1968). On the assumption of typical
Type I behaviour, a distance in excess of 10 kpc is implied in
the absence of any interstellar absorption, (see Fig. 1.5).
While at the relatively high galactic latitude of the source
absorption should not be excessive, some allowance for it must
be made. According to Minkowski (1964), on his interpretation
of the historical colour descriptions the reddening of Kepler's
supernova was between 0.0 and 0.4 magnitudes greater than for
Tycho's supernova, although we have already commented on discre-
pancies in the colour descriptions so that the method could not
be considered reliable. Nevertheless, assuming a total interstellar absorption in the range 1 to 2 magnitudes would place the
supernova at a distance of between 6.5 to 10 kpc. It is impossible
to be more definite, although deceleration of the remnant suggests
that it is expanding into a region of interstellar density
characteristic of the galactic plane despite its large z-distance
(~ 1 kpc). It would therefore seem an attractive hypothesis to
place the supernova in the nuclear-bulge of the galaxy. In
subsequent analysis we have accordingly adopted a distance of
10 kpc (with considerable uncertainty).

A search for the optical remnant of Kepler's supernova by Baade
(1943) revealed a number of bright 'knots' (depicted in Fig. 11.2),
with a few faint filaments. Spectral investigations by Minkowski
(1943) showed emission lines of O I , O III , N II , S II ,
and H . Intercomparison of plates of the nebula taken with the
200 " Mt. Wilson telescope over the past 20 years shows long-term

variations in the brightness of various features, and a radial
movement of the brightest knot at ~0.03" per year (which at the
assumed distance of 10 kpc corresponds to a velocity of ~1400
km sec^{-1}). This velocity is considerably less than the required
average velocity of expansion since the outburst. A similar
discrepancy was previously noted for Tycho's SNR, and the explanation for this suggested in Chapter 10 is presumably applicable
here also. Atempts to identify the stellar remnant of the
outburst have so far been unsuccessful.

Kepler's SNR is the only 'young' remnant on the near-side of the
galaxy which has not yet been detected in X-rays. However this
is hardly surprising in view of its extreme distance placing
it beyond the sensitivity range of the present generation of
X-ray instruments. Additionally, there are several strong X-ray
sources in this region of sky which would produce confusion
problems in any attempted observation.

No supernova has been observed in our galaxy since Kepler's
time. The nearby (distance ~3 kpc) strong non-thermal radio
source Cas A (G111.7-2.1) has the morphological and spectral
characteristics of an SNR, and is believed to be the remnant of a
supernova which occurred a mere 2 to 3 centuries ago. However
there is no historical record which could possibly correspond to
this event. A possible explanation for its non-detection could
be that it occurred in a region of anomalously high absorption
(~2.5 magnitudes per kpc) so that despite its proximity to Earth
it did not reach the brightness required to be readily discovered.
Even allowing for such an effect, the fact that it was not seen
in Europe or the Far East despite being circumpolar remains
something of a mystery.

The next galactic supernova is eagerly awaited by the world's
astronomical fraternity, and the means of observing such an event
are now very powerful indeed. There can be no doubt that a
supernova outburst would precipitate urgent and concentrated
astronomical observation on a scale never before undertaken.
Professor Sidney van den Bergh of the David Dunlap Observatory,
University of Toronto has already circulated plans to astronomers
throughout the world of observational requirements following a
galactic supernova. The immediate impact on modern astronomy
would undoubtedly be considerable - the long-term impact on
civilization on the planet Earth of a particularly nearby supernova
could be extremely dramatic, as we will speculate in Chapter 12.
In the mean time the world's astronomers and astrophysicists
wait, hoping that during their life-time they may be privileged
to witness one of the Universe's greatest spectacles.

Chapter 12

SOME THOUGHTS ON THE EVOLUTION
OF SUPERNOVA REMNANTS

In the preceding chapters we have presented eight proposed associations of historically recorded new stars with SNRs; these
associations, varying from certain to possible, are summarized in
Table 12.1

Table 12.1

Supernova	Radio Remnant	Remarks
AD 185	G315.4-2.3	Probable
AD 386 (?)	G11.2-0.3	Possible
AD 393	G348.5+0.1 or G348.7+0.3	Possible
AD 1006	G327.6+14.5	Certain
AD 1054	G184.6-5.8	Certain
AD 1181	G130.7+3.1	Probable
AD 1572	G120.1+1.4	Certain
AD 1604	G4.5+6.8	Certain

As mentioned in Chapter 1, most of the 120 galactic radio sources
thought to be the remnants of supernovae are probably so old that
no records of the outburst can be expected to exist. Thus the few
identifications of historical supernovae with remnants must provide
valuable observational evidence to test current theories on the
evolution of SNRs.

The most straightforward method of investigating the dynamical
evolution of SNRs is to use the measured diameters and current
expansion velocities of those whose ages are reliably known.
This method has been of limited success in the past because of
the paucity of historical supernova-SNR associations, and the
currently accepted evolutionary picture has been derived chiefly
from theoretical work and indirect measurements. It now seems
worthwhile assessing what new light the six certain or probable
associations given in Table 12.1 cast on the evolution of SNRs.

The picture of the evolution of SNRs which emerges from theoretical investigations was discussed briefly in Chapter 4. After
a few hundred years of near-free expansion, the evolution resembles
that of an adiabatic blast wave created by releasing energy at
a point in a homogeneous gas. The diameter $D(pc)$ of the shock
wave preceding the expanding shell of swept-up interstellar
material is then described by the Sedov (1959) similarity solution:

$$D = 4.3 \times 10^{-11} \left(\frac{E_O}{n} \right)^{\frac{1}{5}} t^{\frac{2}{5}} \tag{1}$$

where t(yr) is the time elapsed since the explosion, E_O (erg) is the energy released in the outburst, and n (cm^{-3}) is the number density of H atoms in the interstellar medium.

The shortage of reliable age calibrators has made the direct investigation of the D - t relationship (1) difficult. For this reason the radio surface brightness Σ, which easily lends itself to observational investigation, has usually been used in evolutionary studies. It is then postulated that the variation of this parameter with linear diameter and time may be approximated by the relationships

$$\Sigma = AD^{\beta} \tag{2}$$

$$\Sigma = Bt^{\gamma} \tag{3}$$

These two expressions may be combined to give the relationship for variation of linear diameter with time

$$D = Ct^{\delta} \tag{4}$$

Because of the difficulty in evaluating the Σ - t and D - t relationships, cumulative distributions are introduced:

$$N(\Sigma) = P\Sigma^{3} \tag{5}$$

where $N(\Sigma)$ is the number of SNRs with surface brightness greater than Σ, and

$$N(D) = QD^{\eta} \tag{6}$$

where $N(D)$ is the number of SNRs with diameter less than D. If a constant rate of supernova outbursts is assumed, then

$$N(\Sigma) = P\Sigma^{3} = \frac{t}{\tau} \qquad \text{and} \qquad N(D) = QD^{\eta} = \frac{t}{\tau}$$

where τ is the characteristic interval between supernova outbursts. Thus the functional dependence of Σ or D on t may be found from the cumulative distributions and only the constant τ requires calibration using SNRs of known age and measured Σ or D. The above relationships have been used in a number of SNR studies, and are shown above in the form adopted by Mills (1974).

The Σ - D relationship derived at a frequency of 408 MHz by Clark and Caswell (1976) is shown as Fig. 12.1.

$$\Sigma_{408} = 10^{-15} \, D_{pc}^{-3}$$

$$\text{for} \ \ \Sigma > 3 \times 10^{-20} \ Wm^{-2} \ Hz^{-1} \ Sr^{-1}$$

and
$$\Sigma_{408} = 3.6 \times 10^{-5} \, D_{pc}^{-10}$$

$$\text{for } \Sigma < 3 \times 10^{-20} \text{ Wm}^{-2} \text{ Hz}^{-1} \text{ Sr}^{-1}$$

It should be emphasized that the Σ - D approach is at present an empirical one with little theoretical foundation, but the quality of present observational data and their statistical implications provide valuable constraints against which future theoretical models may be tested.

The statistical investigation by Clark and Caswell (1976) confirmed that the functional dependence of D on t suggested by the Sedov equation (1) is observed in practice for the majority of SNRs. Additional data, and in particular the frequency of occurrence of supernovae $f = 1/\tau$, are required to determine the constant of proportionality; one method is to estimate the rate of occurrence of supernovae using all those of known ages, another uses the individual sizes and ages of historically recorded supernovae, and a rather different approach utilizes the X-ray data for older supernova remnants.

Fig. 12.1. The surface brightness vs. linear diameter curve of Clark and Caswell (1976). The solid line was determined by the calibrators indicated by filled circles. Those shown with open circles are above the surface brightness range of interest or show anomalous properties. At low surface brightness values the five calibrators indicated by crosses are all of low accuracy, but nevertheless all suggest a steepening of the curve.

From consideration of all three approaches, Clark and Caswell
(1976) concluded that the best estimate of E_0/n was 5×10^{51}
erg cm^3, a somewhat greater value than assumed in most previous
studies.   Clark and Culhane (1976) found a slightly smaller value
from a detailed consideration of the third method alone.   However,
both investigations are consistent with a relationship

$$D \quad 0.9t^{2/5} \tag{7}$$

previously quoted in Chapter 4.

The complete set of evolutionary relationships in the form of
equations (2) to (6) derived by Clark and Caswell (1976) for
$\Sigma > 3 \times 10^{-20}$ Wm^{-2} Hz^{-1} Sr^{-1} is

$$\Sigma = 10^{-15} D^{-3}$$

$$D = 0.9 \ t^{2/5}$$

$$N = 8 \times 10^{-3} D^{5/2}$$

$$N = 2.5 \times 10^{-15} \Sigma^{-5/6}$$

$$\Sigma = 1.25 \times 10^{-15} \ t^{-6/5}$$

Turning now to the individual historical supernovae:
The remnants of Tycho's and Kepler's supernovae have current
linear diameters of 13.8 and 9.3 pc (see Table 4.2) and ages of
403 and 371 years respectively;   inserted in equation (7), these
yield values of E_0/n of 21×10^{51} erg cm^3 for Tycho's supernova
and 3.4×10^{51} erg cm^3 for Kepler's supernova.   Both radio remnants,
showing characteristic peripheral brightening, are of slightly
below average Σ relative to the mean Σ - D relationship.   The
fact that E_0/n appears to be slightly above average for Tycho's
supernova could result from an over estimate of its distance;
however reducing the distance estimate could cause the value of
 Σ to lie somewhat further below the mean Σ - D curve.   Since
the distance seems well established (Goss et al, 1973), at least
as a _lower_ limit, it seems more likely that E_0/n is simply
exceptionally large.

It has previously been noted that the radio remnants for the
supernovae of AD 1181 (3C58) and AD 1054 (the Crab Nebula) are
atypical,in that both show central brightening and flatter than
average non-thermal spectra with significant polarisation confirming a synchrotron origin for the emission.   Because of other
enigmatic differences, the discussion of the Crab Nebula is
deferred until later.   In the case of 3C58, despite not displaying
the peripheral brightening of SNRs, adoption of the outer boundary
of radio emission as the shock front diameter leads to $E_0/n = 2.8 \times 10^{5}$
erg cm^3, in surprisingly good agreement with most other remnants.
In addition, the values of surface brightness and linear diameter
fit well the proposed Σ - D evolutionary track.

Applying the Σ - D relationship to the proposed remnants of the
supernovae of AD 1006 and AD 185 gives distances of 4 and 3.2 kpc

respectively with ages > 7000 years based on the measurement of
radio surface brightness.   Since we believe the historical identifications to be correct, then clearly these remnants are atypical
in either their D-t evolution or their Σ - D evolution.   Since we
have argued that the AD 1006 supernova was only about 1 kpc from
the Earth, and the AD 185 supernova no more than 2 kpc from the
Earth this would indicate that it is the Σ - D evolution which is
atypical;   i.e. the sources are sub-luminous in their radio
surface brightness.   The value for E_0/n for AD 1006 is then
2.5×10^{51} erg cm^3; for AD 185, if 2 kpc is taken as an upper limit
to the distance, then $E_0/n \sim 10^{52}$ erg cm^3.

The above remnant age and diameter data are plotted in Figure 12.2,
together with the relationship $D = 0.9t^{2/5}$, consistent with the
statistical analysis of Clark and Caswell (1976), and the X-ray
data discussed in Clark and Culhane (1976).   The available
data confirm that, with the exception of the Crab Nebula, the
remnants of the historical supernovae follow the evolutionary
track suggested by the Sedov solution.

Returning to the Crab Nebula, this has long been recognised as a
remarkable object, and many of its properties appear to be unique.
Its current expansion velocity is $\sim$ 2000 km sec :   there is no

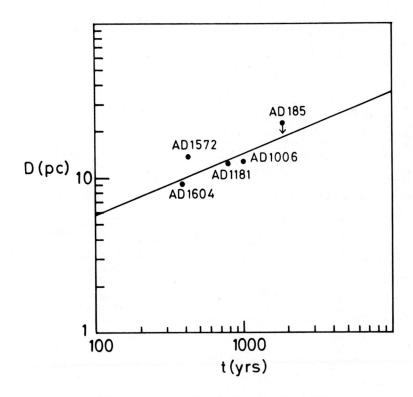

Fig. 12.2.   The D-t relationship

evidence of deceleration, so that if this also represents the initial ejection velocity it is almost an order of magnitude less than
for other supernovae.   The remnant as presently observed at all
wavelengths, with central brightening, clearly owes its existence
to the continued injection of relativistic particles from its
central pulsar, and as outlined in Chapter 8 many of the Crab
Nebula's apparently unique properties may be directly attributed
to the pulsar.   However it is of interest to consider how the
expanding shock wave produced by the supernova outburst may have
evolved, and whether its effects might be observable.

The results discussed earlier suggest that E_0/n typically is
$\sim 5 \times 10^{51}$ erg cm^3.   Adopting this value for the supernova of
AD 1054 gives $D \sim 14$pc from the Sedov relationship, so that for the
usually accepted distance to the Crab Nebula of 2.0 kpc, the
expected angular diameter of the shock wave would be ~ 24 arc min.,
with large uncertainty.   (Note that if the outer boundary of the
observed radio remnant is regarded as the present diameter of the
shock wave, the observed value of E_0/n is only 2×10^{48} erg cm^3.
If one was to then assume a value of $n \sim 0.2$ cm^{-3}, the estimated
initial blast energy $E_0 \sim 10^{47}$ ergs.   This is several orders of
magnitude lower than the value estimated for any other galactic
supernova, and indeed less than the present kinetic energy of
the expanding nebula).

The position of the shock wave delineates a shell-like region of
radio emission in the usually accepted supernova remnant models.
Hazard and Sutton (1971) suggested that the presence of a shell
source of radio emission surrounding the prominent central source
of the Crab Nebula cannot be excluded since it would be very
difficult to detect if its surface brightness were low.

Soft X-ray extensions to the Crab Nebula have been reported by
Toor et al (1975) from lunar occultation data, and by Charles and
Culhane (1976) from observations with the soft X-ray detector on
the satellite Copernicus.   We suggest that these extensions may
be due to the shock from the supernova of AD 1054.   The interstellar
material heated by the shock wave expanding from the site of a
supernova is expected to be a source of X-rays.   Applying the
standard adiabatic shock wave model as summarized by Clark and
Culhane (1976) to the present shock angular diameter of ~ 24 arc min
inferred above for the remnant of AD 1054, and assuming $n \sim 0.2$ cm^{-3}
at the z-distance of the Crab Nebula, gives an estimated intrinsic
X-ray luminosity of $\sim 10^{34}$ erg s^{-1} and an expected observed flux
of $\sim 0.15 \times 10^{-10}$ erg cm^{-2} s^{-1} in the energy interval 0.5 to 1.5
keV.   Such weak emission would not have been detectable above
background with the 2.1 arc min equivalent beamwidth of Copernicus.
We infer therefore that if the soft X-ray extensions actually
detected are evidence of the original shock, they must result
from interaction of the shock with local interstellar density
enhancements - this is the explanation usually invoked for the
"patchiness" of emission for all the X-ray supernova remnants.
The X-ray extensions were observed by the Copernicus satellite to
a diameter of approximately 8 arc min - considerably less than the
24 arc min diameter inferred from the calculations above.   However
the Copernicus observations did not go beyond 5 arc min of the

centre of the source - soft X-ray observations out to 15 arc min
may reveal the total extent of the shock wave.

By suggesting that the present position of the shock wave from
the supernova of AD 1054 lies beyond the region of optical filamentary structure and beyond the outer boundary of the radio
source, many of the apparently anomalous properties of the supernova may be accounted for. Certainly the initial blast energy
could then approach the characteristic value of a few 10^{50} ergs.
There remains the problem of the low proper motion measured for
the optical filaments of the Crab Nebula. The amount of material
producing the optical emission lines through ionisation by the
synchrotron radiation is thought to be at most about one solar
mass (Minkowski, 1968), On the above interpretation for the
supernova of AD 1054 this material is <u>not</u> the initial high-velocity
ejecta of the supernova explosion. (Ejections of different
velocities are recognised as a feature of Type II supernovae).
The present kinetic energy of expansion of the nebula of $\sim 10^{49}$
ergs may have been derived at least in part from the loss of
radial energy of the pulsar.

It appears that there are a number of remnants within the galaxy
that show a general resemblance to the supernova of AD 1054
(Lockhart et al, 1976). It is also possible that the radio source
MSH 15 - 56 is a similar object at a much later stage of evolution
(see Fig. 12.3). Certainly the radio source shows an amorphous
"central" structure like the Crab Nebula; this displays significant polarization indicative of the synchrotron emission process
invoked for supernova remnants, and also has an unusually shallow
radio spectrum. In the case of MSH 15 - 56 however, the present
position of the shock front is clearly delineated by significant
peripheral radio emission lying well beyond the amorphous structure
(Clark et al, 1975).

Apart from the Crab Nebula, the six historical supernovae individually suggest values of E_0/n quite close to the mean value of
5×10^{51} erg cm^3 described in Clark and Caswell (1976) for SNRs
above a given surface brightness threshold. Tycho's SN does
however indicate a somewhat above average value for E_0/n. The
comparisons are of course fairly crude because they depend heavily
on accurate measurements of distance d ($E_0/n \propto d^5$) so that a 15%
error in d results in a factor of 2 error in E_0/n.

In the case of the historical supernovae, a fairly small scatter
in the observed values of E_0/n cannot be due to a systematic
selection effect; furthermore, unless E_0 is proportional to n
(which seems unlikely) it implies that the range of values of
both E_0 and n show quite low dispersions. It has previously
been suggested that n would show quite strong dependence on z,
the perpendicular height above the plane. For example, Clark and
Culhane (1976) assumed a horizontally stratified medium with

$$n(z) = n(0) \exp -\frac{z}{z_0} \tag{8}$$

taking a density on the plane n(0) = 0.5, and a scale height of
~ 120 pc. This dependence on z may have been over estimated

Declination (1950)

Right ascension

Fig. 12.3.   The SNR MSH 15-56.

Consider an interstellar medium comprising both a diffuse gas of
density n_{is} and cloudlets of greatly enhanced density   . A
value of n_{is} (0) as small as 0.2 cm^{-3} may be applicable, and the
scale height of the diffuse component is probably much greater
than that of the cloud/cloudlet component. We would suggest that
the diffuse component should be modelled by equation (8) with a
scale height of, say, z_{is} ~1000pc, with the clouds having mean
density given by

$$n_{cl}(z) = n_{cl}(0) \exp - \frac{z}{z_{cl}}$$

where z_{cl} is the scale height ($\ll z_{is}$) of the clouds. z_{cl} might
be expected to be ~50 pc. If $n_{cl}(0)$ refers to the mean cloud
density averaged over the whole plane, then the variation of
$n_{cl}(z)$ with z-distance could reflect either the spatial variation
of clouds of constant density, or the varying density of individual

clouds occupying a constant fraction of space - or a combination
of these.

Once the SNR is significantly larger than a typical cloud the
shock wave should propagate through the diffuse medium without
being affected significantly by the clouds (McKee and Cowie, 1975).
Indeed Sgro (1975) has shown that a plane shock will completely
recover its form within 2 to 3 cloud diameters of overtaking a
cloud. In the Sedov equation (1) n_{is} should thus be used, and the
dynamical evolution of remnants will then show little dependence
on z up to several hundred pc - within which range most SNRs
are found. Thus for $n_{is} \sim 0.2$ the corresponding typical value of
E_O would be 10^{51} ergs. From general considerations it seems
reasonable that for such a catastrophic event as a supernova,
the threshold might be well defined, with evolution beyond that
threshold leading inevitably to supernova activity showing quite
small dispersions in the explosion parameters.

It has been argued that the radio emission from an SNR is dominated
by quite distinct physical processes at various stages of its
evolution. According to the model of van de Laan (1962), the
interstellar magnetic field compressed by the shock front and
cosmic rays may produce radio emission, but this mechanism is
only satisfactory after the remnant has passed from its adiabatic
phase to a later radiative phase when large compressions are
possible. However since at an adiabatic shock front compression
by only a factor of 4 is obtained, an alternative means of producing the magnetic field is required for the bright radio emission
from young and middle-aged remnants. According to Gull (1973a),
Rayleigh-Taylor instabilities at the interface between the ejecta
and interstellar medium in a young remnant will result in the
formation of a zone of unstable convective mixing. In this zone
random motions will tangle any pre-existing "frozen-in" magnetic
field producing "knots" of greatly enhanced magnetic field
sufficient to describe the strong radio emission from such young
SNRs as Cas A and the remnant of Tycho's SN. After a time this
convection zone expands and dissipates, so that no middle-aged
remnants (age >1000 years) would be expected to be radiating by
this mechanism alone. Thus in a uniform interstellar medium
there is no obvious reason why "middle-aged" SNRs in the adiabatic
phase of their evolution should be strong radio emitters.
Nevertheless, the radio brightness of SNRs appears to decrease
monotonically with time throughout the adiabatic phase, (at least
down to a limiting value of surface brightness - see Fig. 12.1).

McKee and Cowie (1975) suggest that the optical emission of
supernovae in the adiabatic expansion phase comes principally
from interaction of the shock wave with the denser cloudlets of
the interstellar medium, and this is also the mechanism usually
invoked for the patchiness of X-ray emission. Clark and Caswell
(1976) speculated that much of the radio emission may also arise
from this interaction. After the shock has crossed a cloud,
a Rayleigh-Taylor instability will occur (McKee and Cowie, 1975).
This leads to the possibility that in this region for SNRs in the
adiabatic phase non-thermal radio emission may occur due to the
turbulent amplification of the magnetic field and acceleration of

relativistic particles by convective motions, in a similar manner
to the mechanism suggested by Gull (1973a) for young SNRs. If,
as suggested above, the scale height of dense cloudlets, z_{cl},
is significantly smaller than that of the diffuse medium then SNRs
at large values of z may be subluminous in general; however their
rate of expansion, dependent on n_{is}, may not differ significantly
from that of SNRs close to the galactic plane. (Additionally,
SNRs with diameters approaching the scale-height of cloudlets
would be expected to be significantly brighter on their edge
nearest the plane - as in fact is observed for a significant
number of such SNRs. For example, the SNRs shown in Figs.
4.6, 6.4, and 12.3 are all brightest on their edges nearest the
plane.)

The above explanation might account for the particularly subluminous remnants of AD 1006 and AD 1054 (in the latter case ignoring
the emission still being excited by the central pulsar), both at
several cloud scale heights from the plane. Note that such
subluminous remnants would be under-represented in a sample
selected from radio remnant detections relative to an 'historical'
sample (which will in fact favour high latitude objects suffering
less obscuration). It seems clear, however, that at least some
SNRs at quite large values of z are not subluminous in their
radio emission, and in such cases perhaps they occur in regions
with more cloudlets than is common at such z values; indeed
there may be a distinct bias whereby potential supernovae prefer-
entially occur in such regions with quite high concentrations
of dense clouds. Additionally, the longevity of certain SNRs
at large z may result from the presence of an active pulsar
(Clark, 1976).

We finally consider the apparent turnover in the Σ - D relationship noted in Fig. 12.1. Caswell and Clark (1976) slightly revised
the best estimate of this change in slope, proposing that
$\Sigma_{408} = 10^{-15}$ D^{-3} be continued to D = 38pc rather than 32 pc;
at 38pc $\Sigma \sim 1.8 \times 10^{-20}$ Wm^{-2} Hz^{-1}, the age will be $\sim 10,700$ years,
and the expansion velocity still ~ 1000 km s^{-1}. The Σ - D slope
then appears to steepen (to a slope of ~ -10) and since the
expansion velocity is still too high to interpret the slope change
as a transition to the isothermal phase, Caswell and Clark suggested that it is due to a rapid decline in the efficiency of the
radio emission mechanism suggested above when the shock velocity
falls below about 1000 km s^{-1}. In view of the rather speculative
nature of the proposed emission mechanism, this suggestion is not
an obvious conclusion but rather, it is hoped, another clue
to understanding the factors determining the radio intensity.

On this interpretation, the adiabatic expansion of the shock
continues beyond the diameter at which the radio SNRs essentially
fade below the limit of detectability, i.e. to $D \sim 65$ pc. Thus
if D can be estimated by any means (including the use of the Σ - D
curve after its turnover to the steep slope of -10), then applic-
ation of the D-t relationship will probably give a reasonably
good estimate of age.

From the above discussion, it seems likely that most of the radio
SNRs may be regarded as similar objects following a common dynamical
evolutionary track. Dense cloudlets in the interstellar medium
may play an important role in the radio brightness of a particular remnant, although variations in the environment (of a
magnitude commonly encountered) have only slight effects on the
dynamical evolution.

At the beginning of this book we included a quotation from
Zwicky (1965): "The investigation of the remnants of supernovae
and their relation to historical records, both written and unwritten, will be one of the most fascinating tasks awaiting the next
generation of astronomers." The quotation goes on - "These
records may also include various features of a geological and
a biological nature, refering back to evolutionary phenomena of
the Earth's surface and the life on it. Already a book could be
written about the subject of the remnants of supernovae and about
the efforts which have been made to elucidate their characteristics.".

In writing this present book we have limited ourselves to the
written historical records of supernovae, simply because this is
our own particular field of expertise, experience and interest.
We have certainly found, as Zwicky predicted, the detailed
investigation of the written records "fascinating". The undoubtedly equally fascinating task of investigating the unwritten
records must await the attention of others. All we will do here
is to speculate on the possible scope of such investigations, and
the possible effects of a "nearby" supernova on the Earth and its
environment.

McCrea (1975a, b) has proposed that the encounter of the solar
system with a dense cloud of interstellar material during its
passage through a spiral arm of the Galaxy may produce such
climatic catastrophes on Earth as the ice epochs. His thesis,
an extension of several earlier investigations is based on the
expected affect on the solar constant of an increased accretion
rate. Unfortunately the cloud density necessary to produce the
required variation is $10^5 - 10^7$ cm^{-3}, and although clouds with such
extreme densities are thought to exist, Begelman and Rees (1976)
pointed out that they are so compact and unusual as to reduce the
likelihood of the Sun ever encountering one. Begelman and Rees
in fact showed that a much more modest cloud density ($10^2 - 10^3$
cm^{-3}) would prevent the solar wind from reaching the Earth, with
resulting modification of the near-Earth environment; however
the climatic consequences of such a modification are not well
understood. We present here an alternative hypothesis which still
retains McCrea's association with the passage of the solar system
through a spiral arm, but which relates the initiation of an ice
epoch, plus biological catastrophes, to an encounter of the
solar system with a nearby supernova outburst rather than a super-
dense interstellar cloud.

The spiral structure of our galaxy indicates the regions where
bright, short-lived stars illuminate the inter-stellar gas and
dust. The two principal spiral arms are believed to reflect a

self-sustaining density-wave pattern through which the disk of
stars and gas move. Observations indicate dust lanes along the
inner (concave) edge of each arm, and this is interpreted as a
shock-wave effect whereby clouds become temporarily compressed
as they cross the lane. A cloud of interstellar matter in its
orbit round the Galaxy is liable to enter a spiral arm via its
compression lane about once in $\sim 10^8$ y. In general the cloud
material will re-expand on emerging from the lane after $\sim 10^6$ y
(the cloud not necessarily retaining its original identity). But
sometimes a cloud is compressed to a condition for star formation,
and the result is a new galactic cluster which proceeds to traverse the associated spiral arm itself, taking $\sim 10^7$ y to do so.
If this cluster contains a star of mass some 10 $M_\odot$ or greater
it would be expected to evolve rapidly over less than $\sim 10^7$ y,
terminating in a supernova outburst of Type II before emerging
from the spiral arm in which it is formed. In this view supernovae of Type II would occur only in spiral arms (or near galactic
nuclei), in accord with observation. Indeed a "supernova zone"
might be defined lying beyond the compression lane of a spiral
arm;   such a zone is depicted diagrammatically in Fig. 12.4, on
the assumption of a 20 $M_\odot$ stellar life-time of $\sim 5 \times 10^6$ y and
a transverse velocity (at about the solar distance from the
galactic centre) of 200 km/s giving a zone width of ~ 1 kpc.

Clark and Caswell (1976) have concluded from a statistical analysis
of the radio remnants of supernovae that the average interval
between outbursts leaving long-lived observable remnants (i.e.
those lying within spiral arms and evolving only slowly) is 100
years. Shklovsky (1968) prefers an interval of a mere 30 years
for Type I (Population 2) events, but accepts a longer interval
of ~ 100 years for those of Type II (Population 1). Results from
other investigators range between these extremes. In the absence
of a more definite estimate, we will take an average interval
between Type II supernovae of 100 y.

If there is one Type II supernova per 100 y, and if, as seems
reasonable, this is localized to one of the "supernova zones"
proposed above, then this implies 5×10^4 such supernovae in one
(of two) spiral arms in a period of $\sim 10^7$ y (the time for the Sun
to traverse a spiral arm). Since a spiral arm is $\sim 2 \times 10^4$ pc
long, this suggests that 50 supernovae will occur in a 20 pc
strip of a spiral arm's "supernova zone" each time the Sun crosses
it. Thus at least one supernova would be expected to occur within
10 pc of the Sun each time it crosses a spiral arm (i.e about
once every 10^8 years). Note that this simple calculation takes
no account of the thickness of a spiral arm perpendicular to the
plane; however since the bulk of supernova remnants are concentrated to within a few tens of pc of the plane, this simplificat-
ion is not expected to increase the above estimate for the interval between nearby supernovae by more than a factor of two.

A similar estimate is obtained from an interpretation of the
galactic distribution of supernova remnants. Most supernovae
appear to be restricted to within little more than a solar distance of the galactic centre and close to the plane i.e. to a
volume of $\sim 10^{10}$ cubic pc. Making the simplifying assumption that

Fig. 12.4. A schematic representation of part
of a spiral arm, showing a "supernova zone".

the rate of supernovae per unit volume is uniform throughout the Galaxy, then the interval between nearby supernovae occurring within distance R of the Sun is

$$T = \frac{\tau \ 10^{10}}{4/3 \ \pi \ R^3} \quad y$$

where τ y is the characteristic interval between supernovae. With $R \sim 10$ pc, $T \sim 2.5 \times 10^8$ y - of the same order as the value obtained above. One must thus conclude that during the history of our solar system ($\sim 5 \times 10^9$ y) more than 20 supernovae would be expected to have occurred within a distance of 10 pc, and over 2500 within a distance of 50 pc of the Sun.

Estimates of the radiation flash (see for example Colgate, 1968) emitted (mainly in γ and X-rays) in a supernova explosion vary from $\sim 10^{47}$ to 10^{50} ergs. (This represents a very small fraction of the total energy of the outburst of $\sim 10^{52}$ ergs). Assuming the lower value, the peak integrated energy flux expected above the Earth's atmosphere from a supernova at ~ 10 pc distance is 10^6 erg cm^{-2} - this compares with the solar flux of ~ 20 erg/cm^2 sec in the wavelength range of interest. Absorption in the upper atmosphere means that this flash gives no significant surface radiation dose initially. Nevertheless Ruderman (1974) has predicted catastrophic effects on the Earth's protective ozone layer. The ionizing radiation from the supernova flash produces excited free nitrogen atoms in the atmosphere which quickly oxidize. The oxides of nitrogen catalytically destroy ozone, and Ruderman claims that by this mechanism the ozone layer density may be depleted to just a few percent for the order of a hundred years. Such destruction of the ozone layer would expose the Earth's surface, and more particularly life on it during the past 600 million years, to lethal solar ultraviolet radiation.

The expanding shell of ejecta from a supernova at 10 pc distance would reach the Earth after about 2000 years, with the passage of the shell taking several hundred years. During this latter period the density of primary cosmic rays would be expected to be enhanced by perhaps a factor of a hundred (Shklovsky, 1968). About one third of the background radio-activity level at the Earth's surface is contributed by cosmic radiation (the remainder being due to terrestrial factors). Thus a hundredfold increase in cosmic ray intensity would produce a drastic thirty-fold increase in mean level of radioactivity. In addition, the enhanced cosmic ray flux would be expected to produce the same destruction of the ozone layer as produced by the initial radiation flash. The possible biological consequences of such dramatic increases in background radioactivity and ultraviolet radiation fluxes expected for protracted periods sometime after a nearby supernova outburst may have been catastrophic for certain species. Ruderman has suggested that in man, the hugh increase in ultraviolet radiation would give rise to a vast increase in the incidence of skin cancer; in animals with calcified internal skeletons the increased vitamin D production might have been toxic.

The supernova-initiated destruction of the ozone layer would be
expected to have climatic as well as biological consequences. The
long-term reduction of ozone density to possibly just a few percent
of its normal value would lead to an increase of several percent
in solar energy transmitted to the ground, initially producing an
increase of several degrees in mean temperature. (It is interesting to note how many of the historical records of the AD 1006
supernova, the closest to the solar system in historical times,
refer to it being followed by drought, famine, and pestilence.
However at a distance of ~1 kpc the expected energy flux at the
Earth would be only one ten thousanth that of a supernova at
10 pc.) The subsequent increased precipitation and cloud cover
may eventually result in the reflection of an increased proportion
of incident energy and a lowering of mean temperature at the
Earth's surface. Could nearby (closer than 10 pc) supernovae,
occurring on average every few hundred million years, have
initiated the climatic conditions leading to the onset of the ice-
ages on Earth? We believe that this is a possibility worthy of
further consideration.

Possible geological records of nearby supernovae are difficult to
envisage. The increased primary cosmic ray flux during the passage of a supernova remnant shell may have sufficiently modified the
Earth's radiation belts and consequently the geomagnetic field
to be revealed in palaeomagnetic data. Energetic cosmic ray
protons might be expected to produce proton-rich isotopes by the
so-called (p,n) reaction (the ejection of a neutron from an atomic
nucleus following bombardment by a proton), which may show up
for certain epochs in core samples from polar regions; (at middle
and low latitudes the Earth's atmosphere and magnetic field absorb
or deflect nearly all primary cosmic rays.) A more likely site
where such records might be preserved would be a still-active
planet with negligible atmosphere or magnetic field, and Mars
is the obvious example.

The final, and what must be the ultimate hypothesis relating
to the consequences of supernovae must be left to Shklovsky
(1968): "We can imagine that a high level of radioactivity,
caused by some cosmic events which took place in an epoch some
10^9 years ago, may have stimulated the formation of highly complex
compounds from simple organic compounds, and life on the Earth
may have developed on the basis of these complexes."

Perhaps we have wandered too far into the realms of speculation.
We will conclude our present study with a final consideration of
the ancient written records of supernovae.

Astronomy has reaped enormous benefit from the literary traditions
of China, and its richly recorded history. The Far-Eastern
astronomical records represent a unique collection of data from a
full two millenia, and the essential reliability of the observations (despite political, social, and astrological pressures)
are beyond dispute. Duplication of certain observations from the
Arab lands after about the 9th century AD provide useful confirmat-
ory results. The post-Renaissance contributions from Europe may

have surpassed the ancient observations in precision , but not in
their impact on modern astrophysical investigations. We can
only reiterate and reinforce the claim we made at the end of
Chapter 1, that there can be little doubt that the historical
astronomical records must be regarded as among the most valuable
legacies which the ancient world has bequeathed to modern
science.

REFERENCES

The following abbreviations are used for frequently occurring reference sources :-

A.A.	Astronomy and Astrophysics
A.J.	Astronomical Journal
A.J.P.	Australian Journal of Physics
A.N.	Astronomische Nächrichten
A.P.A.E.	Astronomical Papers of the American Ephemeris
Ap. J.	Astronomical Journal
Ap. Letters	Astrophysical Letters
M.R.A.S.	Memoirs of the Royal Astronomical Society
M.N.R.A.S.	Monthly Notices of the Royal Astronomical Society
P.A.S.P.	Publications of the Astronomical Society of the Pacific
Q.J.R.A.S.	Quarterly Journal of the Royal Astronomical Society

Allen, C.W., 1973. Astrophysical Quantities (3rd Edition) (Univ. of London Press).

Argelander, F., 1864. A.N., 62, 273.

Arnold, T., 1885. Simeonis monachi Dunelmensis Historia Regum (HMSO, London).

Baade, W., 1942. Ap. J., 96, 188.

Baade, W., 1943. Ap. J., 97, 119.

Baade, W., 1945. Ap. J., 102, 309.

Baldwin, J.E., 1971. In The Crab Nebula (Reidel, Holland).

Barbon, R., Ciatti, F., and Rosino, L., 1974 a and b. In Supernova and Supernova Remnants (Reidel, Holland).

Bemporad, A., 1904. Mitt. Heidelberg, No. 4.

Bielenstein, H., 1950. Bull. Mus. Far. E. Antiq., 22, 127.

Biot, E., 1843. Connaissance des Temps., Additions, p60.

Böhme, S., 1936. A.N., 262, 479.

Bolton, J.G., 1948. Nature, 162, 141.

Bolton, J.G., and Stanley, G.J., 1949. Australian. J. Sci. Res., 2A, 139.

Bowyer, C.S., et al., 1964. Science, 146, 912.

Brandt, J.C., et al., 1975. In Archaeoastronomy in Pre-Columbian America. ed. A.F. Aveni. (Univ. of Texas Press).

Brosche, P., 1967. In Variable Stars. IAU Info. Bull. No. 192.

Caswell, J.L., 1967. M.N.R.A.S., _136_, 11.
Caswell, J.L., 1970. A.J.P., _23_, 105.
Caswell, J.L., and Clark, D.H., 1975. A.J.P. Astrophys. Suppl.,
        No. _37_, 57.
Caswell, J.L., Clark, D.H., and Crawford, D.F., 1975. A.J.P.
        Astrophys. Suppl., No. _37_, 39.
Caswell, J.L., et al., 1976. A.A. - in press.
Caswell, J.L., and Clark, D.H., 1976. M.N.R.A.S. - in press.
Chambers, G.F., 1909. The Story of Comets. (Clarendon Press,
        Oxford.)
Charles, P.A., and Culhane, J.L., 1975. Scientific American,
        _233_, 38.

Ch'en-yuan, 1935. Sung-hui-yao-chi-kao (Peking).
Chevalier, R.A., 1974. Ap. J., _188_, 501.
Chu Sun-il, 1968. J. Korean Astr. Soc., _1_, 29.
Clark, D.H., 1976. M.N.R.A.S., _175_, 77p.
Clark, D.H., Caswell, J.L., and Green, A., 1973. Nature, _246_, 28.
Clark, D.H., Caswell, J.L., and Green, A., 1975. A.J.P. Astrophys.
        Suppl., No. _37_, 1.
Clark, D.H., Green, A., and Caswell, J.L., 1975. A.J.P. Astrophys.
        Suppl. No. _37_, 75.
Clark, D.H., and Caswell, J.L., 1976. M.N.R.A.S., _174_, 267.
Clark, D.H., and Culhane, J.L., 1976. M.N.R.A.S., _175_, 573.
Clark, D.H., and Stephenson, F.R., 1976. Q.J.R.A.S., _17_, 290.
Davison, K., and Tucker, W., 1970. Ap. J., _161_, 437.
Davison, P.J.N., Culhane, J.L., and Mitchell, R.J., 1976. Ap. J.,
        _206_, L37.
Day, G.A., Caswell, J.L., and Cooke, D.J., 1972. A.J.P. Astrophys.
        Suppl., No. _25_, 1.
Dickel, J.R., McGuire, J.P., and Yang, K.S., 1965. Ap. J., _142_,
        798.
Dickel, J.R., and Yang, K.S., 1965. Ap. J., _142_, 1642.
Dickel, J.R., and McKinley, R.R., 1969. Ap. J., _155_, 67.
Dickel, J.R., Milne, D.K., Kerr, A.R., and Ables, J.G., 1973.
        A.J.P., _26_, 379.
Dicks, D.R., 1960. The Geographical Fragments of Hipparchus.
        (Univ. of London Press.)
Dombrovsky, V.A., 1954. Akad. Nauk. USSR., _94_, 1021.
Downes, D., 1971. A.J., _76_, 305.
Dreyer, J.L.E., 1890. Tycho Brahe (Adam and Charles Black,
        Edinburgh).
Dreyer, J.L.E., 1906. History of the Planetary System from
        Thales to Kepler. (Cambridge Univ. Press).
Duin, R.M., and Strom, R.G., 1975. A.A., _39_, 33.
Duin, R.M., et al., 1975. A.A., _38_, 461.
Duncan, J.C., 1921. Proc. Nat. Acad. Sci., _7_, 179.
Duncan, J.C., 1939. Ap. J., _89_, 482.
Duyvendak, J.J.L., 1942. P.A.S.P., _54_, 91.
Ellis, F.H., 1975, In Archaeoastronomy in Pre-Columbian America.
        ed. A.F. Aveni. (Univ. of Texas Press).
Ewen, H.I., and Purcell, E.M., 1951. Nature, _168_, 356.
Fessenkov, V.G., 1962. In Physics and Astronomy of the Moon.
        ed. Z. Kopal. (Academic Press, New York).
Fotheringham, J.K., 1919. M.N.R.A.S., _79_, 162.

Friedman, H., Byram, E.T., and Chubb, T.A., 1967. Science, 156,
        374.
Goldstein, B.R., 1965. A.J., 70, 105.
Goldstein, B.R., 1966. Nature, 211, 504.
Goldstein, B.R., and Ho Peng Yoke, 1965. A.J., 70, 748.
Goss, W.M., and Schwarz, U.J., 1971. Nature, 234, 52.
Goss, W.M., et al., 1972. Ap. J. Suppl., 24, 123.
Goss, W.M., Schwarz, U.J., and Wesselius, P.R., 1973. A.A., 28,
        305.
Green, A.J., 1971. A.J.P., 24, 773.
Green, A.J., 1974. A.A. Suppl., 18, 325.
Gull, S.F., 1973a. M.N.R.A.S., 161, 47.
Gull, S.F., 1973b. M.N.R.A.S., 162, 135.
Gull, S.F., 1975. M.N.R.A.S., 171, 237.
Hagen, J.P., and McClain, E.F., 1954. Ap. J., 120, 368.
Han Yu-shan, 1955. Elements of Chinese Historiography.
        (W.M. Hawley, Hollywood).
Hanbury-Brown, R., and Hazard, C., 1952. Nature, 170, 364.
Haslam, C.G.T., and Salter, C.J., 1971. M.N.R.A.S., 151, 385.
Hawkins, F.J., et al., 1974. M.N.R.A.S., 169, 41.
Hazard, C., and Sutton, J., 1971. Ap. Letters, 7, 179.
Hellman, C.D., 1960. Isis, 51, 322.
Henning, K., and Wendker, H.J., 1975. A.A., 44, 91.
Herman, B.R., and Dickel, J.R., 1973. A.J., 78, 879.
Hewish, A., and Okoye, S.E., 1964. Nature, 203, 171.
Higgs, L.A., and Halperin, W., 1968. M.N.R.A.S., 141, 209.
Hill, E.R., 1964. Symp. IAU-URSI, No. 20., 107.
Hill, E.R., 1967. A.J.P., 20, 297.
Hill, I.E., 1972. M.N.R.A.S., 157, 419.
Ho Peng Yoke, 1962. Vistas in Astronomy, 5, 127.
Ho Peng Yoke. 1966. The Astronomical Chapters of the Chin Shu.
        (Mouton, Paris.)
Ho Peng Yoke, 1969. J. Asian History, 3, 135.
Ho Peng Yoke, 1970. Oriens Extremus, 17, 63.
Ho Peng Yoke, Paar, T.H., and Parson, P.W., 1970. Vistas in
        Astronomy, 13, 1.
Hsi Tsê-tsung, 1955. Acta Astr. Sinica, 3, 183.
        Trans. 1958 in Smithson. Contr. Astrophys., 2, 109.
Hsi Tsê-tsung, and Po Shu-jen, 1965. Acta. Astr. Sinica, 13, 1.
        Trans. NASA Tech. Trans. TTF-388. Abridged trans. by
        K.S. Yang, 1966 in Science, 154, 597.
Hsüeh Chung-san and Ou-yang, I, 1956. A Sino-Western Calendar for
        Two Thousand Years, AD 1-2000, (Peking).
Hughes, M.P., Thompson, A.P., and Coluin, R.S., 1971. Ap. J.
        Suppl., 23, 323.
Humboldt, A. von., 1851. Cosmos. (English trans. by E.C. Otte),
        vol. III, (Bohn, London).
Ilovaisky, S.A., and Lequeux, J., 1972. A.A., 18, 169.
Isles, J.E., 1976. J. Brit. Astr. Assn., 86, 245.
Jao Tsung-i, 1959. Yin-tai-cheng-pu-jen-wu-t'ung-kao
        (Hong Kong Univ. Press).
Jenkins, L.F., 1952. General Catalogue of Trigonometric Stellar
        Parallaxes. (Yale Univ. Observatory, New Haven).
Jones, J., Spencer, 1961. General Astronomy, (4th Edition),
        (Edward Arnold Ltd., London).

Jones, B.B., 1973. A.J.P., 26, 545.
Kanda Shigeru, 1935. Nihon Temmon Shiryo. (Koseisha, Tokyo)
Keen, N.J., et al., 1973. A.A., 28, 197.
Kerr, A.R., 1968. In Nebulae and Interstellar Matter, Vol. VII,
        of Stars and Stellar Systems. (Chicago).

Kestenbaum, H., Angel, J.P.P., and Novick, R., 1971. Ap. J.,
        164, L87.
Kiang, T., 1972. M.R.A.S., 76, 27.
Kin Yong-woon, 1974. Korea Journal, 14, 4.
Kukarkin, B.V., et al., 1969. General Catalogue of Variable
        Stars. (Moscow).
Kundu, M.R., 1971. Ap. J., 165, L55.
Kundu, M.R., and Velusamy, T., 1972. A.A., 20, 237.
Lundmark, K., 1921. P.A.S.P., 33, 225.
Lynn, W.T., 1884. Observatory, 7, 17 and 75.
Mayall, N.U., 1937. P.A.S.P., 49, 104.
Mayall, N.U., and Oort, J.H., 1942. P.A.S.P., 54, 95.
Mayer, C.H.T., McCullough, T.P., and Sloanaker, R.M., 1957.
        Ap. J., 126, 468.
McKee, C.F., 1974. Ap. J., 188, 335.M
McKee, C.F., and Cowie, L.L., 1975. Ap. J., 195, 715.
Menon, T.K., and Williams, D.R.W., 1966. A.J., 71, 392.
Miller, W.C., 1955a. Plateau, Mus. of N. Arizona, 27, 6.
Miller, W.C., 1955b. Astron. Soc. Pacific, Leaflet No, 314.
Mills, B.Y., Slee, O.B., and Hill, E.R., 1957. A.J.P., 11, 360.
Mills, B.Y., 1974. In Galactic Radio Astronomy. (Reidel, Holland).
Milne, D.K., 1969. A.J.P., 22, 613.
Milne, D.K., 1970. A.J.P., 23, 425.
Milne, D.K., 1971a. A.J.P., 24, 429.
Milne, D.K., 1971b. A.J.P., 24, 757.
Milne, D.K., 1972. Ap. Letters, 11, 167.
Milne, D.K., and Hill, E.R., 1969. A.J.P., 22, 211.
Milne, D.K., and Dickel, J.R., 1975. A.J.P., 28, 209.
Minkowski, R., 1942. Ap. J., 96, 199.
Minkowski, R., 1943. Ap. J., 97, 128.
Minkowski, R., 1959. Paris Symposium on Radio Astronomy.
        (Stanford Univ. Press).
Minkowski, R., 1964. Ann. Rev. Astr. Ap., 2, 247.
Minkowski, R., 1966. A.J., 71, 371.
Minkowski, R., 1968. In Nebulae and Interstellar Matter, Vol. VII.
        of Stars and Stellar Systems. (Chicago).
Minkowski, R., 1971. In The Crab Nebula. (Reidel, Holland).
Morohashi Tetsuji, 1955. Dai Kanwa Jiten (Tokyo).
Moule, A.C., and Yetts, W.P., 1957. The Rulers of China.
        (Routledge and Kegan Paul, London).
Muller, P.M., and Stephenson, F.R., 1975. In Growth Rhythms and
        the History of the Earth's Rotation.
        (John Wiley and Sons Ltd., London).
Muratori, L.A. (ed) 1723. Rerum Italicarum Scriptores, 25 vols.
        (Milan).
Needham, J., 1959. Science and Civilisation in China, vol. 3.
        (Cambridge Univ. Press).
Needham, J., Wang-ling, and Price, D. de S., 1960. Heavenly
        Clockwork, (Cambridge Univ. Press).
Newcomb, S., 1895. A.P.A.E., 6.

Newcomb, S., 1910. A.P.A.E., 9.
Newton, R.R., 1972. Medieval Chronicles and the Rotation of the
        Earth. (The John Hopkins Univ. Press).
Oppolzer, T.R., von, 1887. Canon der Finsternisse.
        (Kaiserlichen Akadamie der Wissenschaften, Wien).
Payne - Gaposchkin, Co, 1956. Introduction to Astronomy.
        (Eyre and Spottiswoode, London).
Payne - Gaposchkin, C., 1957. The Galactic Novae.
        (North Holland Publishing Co., Amsterdam).
Pertz, G.H., (ed)., 1826. Monumenta Germaniae Historica Scriptores,
        32 vols. (Hahn, Hanover).
Pingré, A.G., 1783. Cometographie ou Traite Historique et
        Theorie des Comets, vol. 1 (Paris).
Porter, N.A., 1974. J. Hist.Astr., 5, 99.
Pskovskii, Y.P., 1963. Sov. Astr., 7, 501.
Pskovskii, Y.P., 1972. Sov. Astr., 16, 23.
Radhakrishnan, V., et al., 1972. Ap. J. Suppl., 24, 49.
Rigge, W.F., 1915. Pop. Astr., 23, 29.
Rosenberg, I., 1970. M.N.R.A.S., 151, 109.
Roger, R.S., Bridle, A.H., Costain, C.H., 1973. A.J., 78, 1030.
Rosenberg, I., and Scheur, P.A.G., 1973. M.N.R.A.S., 161, 27.
Ruderman, M.A., 1974. Science, 184, 1079.
Rufus, W.C., 1913. J. Roy. Asiatic Soc. (Korea Branch), 4, 23.
Rufus, W.C., and Chao, C., 1944. Isis, 35, 316.
Rufus, W.C., and Tien Hsing-chih, 1945. The Soochow Astronomical
        Chart. (Univ. of Michigan Press, An Arbor.).
Schlier, O., 1934. A.N., 254, 181.
Schlesinger, F., 1935. General Catalogue of Stellar Parallaxes
        (Yale Univ. Observatory, New Haven).
Schlesinger, F., 1940. Catalogue of Bright Stars, (Yale Univ.
        Observatory, New Haven).
Schoenberg, E., 1929. In Handbuch der Astrophysik II.
Sedov, L.I., 1959. Similarity and Dimensional Methods in Mechanics,
        (Academic Press, New York).
Seward, F., et al., 1976. Ap. J., 205, 238.
Sgro., A.G., 1975. Ap. J., 197, 621.
Shower, P.A., 1969. M.N.R.A.S., 142, 273.
Shaver, P.A., and Goss, W.M., 1970. A.J.P., Astrophys. Supple.,
        14, 77.
Shinoda Minoru, 1960. The Founding of the Kamakura Shogunate
        1180 - 1185. (Columbia Univ. Press, New York).
Shklovsky, I.S., 1953. D.A.N., 90, 983.
Shklovsky, I.S., 1960a. Sov. Astr., 4, 243.
Shklovsky, I.S., 1960b. Sov. Astr., 4, 355.
Shklovsky, I.S., 1962. Sov. Astr., 6, 162.
Shklovsky, I.S., 1966. Sov. Astr., 10, 6.
Shklovsky, I.S., 1968. Supernovae, (John Wiley and Sons Ltd.,
        London).
Shklovsky, I.S., 1974. Sov. Astr., 18, 1.
Slee, O.B., and Dulk, G.A., 1974. In Galactic Radioastronomy
        (Reidel, Holland).
Staelin, D.H., and Reifenstein, E.C., 1968. Science, 162, 1481.
Strom, R.G., and Duin, R.M., 1973. A.A., 25, 351.
Stephenson, F.R., 1971. Q.J.R.A.S., 12, 10.

Stephenson, F.R., 1975.   In Origin of Cosmic Rays, (Reidel,
        Holland).
Stothers, R., 1976. In preparation.
Tammann, G.A., 1966. Nature, 210, 511.
Tchang, M., 1905. Varietes Sinologiques, 24. (Shanghai).
Toor, A., et al., 1976. Ap. J. In press.
Tuckerman, B., 1964. Mem. Amer. Phil. Soc., 59.
Trimble, V., 1968. A.J., 73, 535.
Trimble, V., 1971. In The Crab Nebula, (Reidel, Holland).
Tsui Chi, 1947. A Short History of Chinese Civilisation,
        (Victor Gollancz, London).
Van den Bergh, S., 1970. Nature, 225, 502.
Van den Bergh, S., Marscher, A.P., and Terzian, Y., 1973.
        Ap. J.,Suppl., 26, 19.
Van der Laan, S., 1962. M.N.R.A.S., 124, 179.
Velusamy, T., and Kundu, M.R., 1974. A.A., 32, 375.
Vettolani, G., and Zamorani, G., 1976. In preparation.
Weiler, K.W., and Seielstad, G.A., 1971. Ap. J., 163, 455.
Westerlund, B.E., 1969. A.J., 74, 879.
Whiteoak, J.B., and Gardner, F.F., 1968. Ap. J., 154, 807.
Williams, D.R.W., 1973. A.A., 28, 309.
Williams, J., 1871. Chinese Observations of Comets, (London).
Willis, A.G., 1973. A.A., 26, 237.
Wilson, R.A., 1971. In The Crab Nebula (Reidel, Holland).
Wilson, T.L., 1970. Ap. Lett., 1, 95.
Winkler, P.F., and Laird, F.N., 1976. Ap. J. In press.
Woltjer, L., 1964. Ap. J., 146, 1309.
Woltjer, L., 1970. In Interstellar Gas Dynamics, (Reidel, Holland).
Woltjer, L., 1971. In The Crab Nebula, (Reidel, Holland).
Woltjer, L., 1972. Ann. Rev. Astr. Astrophys., 10, 129.
Yabuuchi Kiyoshi, 1967. Sogen Jidai no Kagadu Gijutsu Shi, (Kyoto).
Yang, K.S., 1966. Science, 154, 597.
Zwicky, F., 1939. Phys. Rev., 55, 986.
Zwicky, F., 1965. In Stellar Structure, Vol. VIII of Stars and
        Stellar Systems, (Chicago)

INDEX

accuracy
  of Arabic measurements 127.
  of European measurements
      172, 180 ff, 202.
  of Far Eastern measurements
      34, 129-130, 179, 194.
  of Oriental dates 86, 104-
      105.
adiabatic expansion phase 63.
Alī ibn Ridwān 120 ff.
Altobelli, I. 196.
American Indians 16.
Antares 195.
Arabic observations 7, 18,
      114 ff.
Aristotle 5, 17, 172.
armillary spheres 144-145,
      148.
asterisms, oriental 33 ff,
      and passim.
astrology 25, 28, 33-34, 85,
      102-104, 194.
Astronomical Bureau 25, 27,
      129, 131, 142.
astronomical day 168.
atmospheric absorption
      (extinction) 94-95, 106,
      109, 133-134, 200.
auspicious stars 115-117.

calendar, Chinese 25, 29 ff.
Cassiopeia 162 ff, 172 ff.
Cas A 60, 70, 81, 169, 206.
catalogue
  of pre-telescope novae
  and supernovae 43 ff.
  of supernova remnants 64 ff.
  of temporary stars 41 ff.
cave paintings (rock art) 16.
Ch'ien-k'ang (Nan-ching) 103.
China, history of 19 ff.
Chinese
  characters 115, 193.
  observations 25 ff, and
      passim.
Classics, Greek and Roman 14.
climatic catastrophies 221.
collapse, stellar 2, 3.
comets 40 ff, 84, 120, 123-124,
      147-148, 155, 173.
comparison stars 176 ff, 200 ff.
continuum radio emission 57.
Copernicus, Nicholas 172.
copyists errors 144.
cosmic rays 2.
cosmology, medieval European
      172.
Crab Nebula 7, 59, 61, 140 ff,
      212, 213.
cross staff 185.

Babylonian tablets 14-15.
Benevento (Italy) 122.
binary systems, close 4.
biological catastrophies 220,
      221.
black holes 2, 73.
Brahe, Tycho 6, 19, 133, 163,
      172 ff, 191, 196, 199.
broom stars 40 ff, 119, 148,
      176.
burning of the books 5, 21.
bushy stars: see rayed stars.

date conversion 33.
dating errors 87, 94, 105-106,
      108-109, 169.
daylight
  comets 147.
  visibility 125, 135, 140,
      142, 147, 150, 152, 175,
      177.
Digges, Thomas 185-186.
divination 19, 25, 34, 119.
double-lobe structure, of super-
      nova remnants 70-71.

duration of visibility   50 ff,
    and passim.
dynastic histories   5, 21 ff.
dynasties, Chinese   19 ff.

Hipparchus   14.
Historiography, Bureau of   21,
    24, 142, 152.
horizon profiles   94, 127-128.

eclipses   15, 19, 31, 142-143.
ecliptic   106, 144, 153.
errors of measurement, Tycho's
    180 ff.
European observations   7, 16-
    19, 114 ff, 172 ff, 191 ff.
evolution
    of stars   1-4.
    of supernova remnants   13,
    63, 70.
extinction phase   63.

ibn Ridwān, Alī   120 ff.
ice epochs   217.
imperial colours   117, 142.
inch, Chinese   148-149.
intercalary months   103.
interstellar absorption   8-9, 53,
    109, 135, 152, 189.
ionised hydrogen clouds   57-59,
    74.
isothermal expansion phase   63.

falsification (fabrication) of
    observations   26-27, 85-86,
    104.
Far Eastern observations   5,
    19, 25 ff, 40 ff, and
    passim.
Faraday rotation   158.
flux density, radio   63, and
    passim.
free-expansion phase   63.
free-free interaction   57-58.
Fujiwara Sadaie   41, 118, 130-
    131, 143, 161.
fusion, nuclear   1, 2.

Japan, history of   27-28.
Japanese observations   28, 42-43,
    114 ff, 140, 143-145, 147,
    161-163.
Jupiter   16-17, 85-87, 168.

K'ai-fêng (China),   128, 132.
Kepler, Johannes   19, 174, 191 ff.
Kepler's supernova   7, 59, 163,
    169, 191 ff.
Korea, history of   28-29.
Korean
    alphabet   29.
    observations   29, 119-120, 140,
    162, 167, 169, 176, 191 ff.
Kyōto   27, 128.

galactic
    coordinates   51 ff.
    equator   106.
geological consequences of
    supernovae   221.
guest stars   40 ff, and passim.

Liege (Belgium)   123.
light curves of new stars   179-
    180, 202.
limb-brightening   59.
limit, naked eye   10, 165.
line emission   57, 155, 189.
Lobbes (Belgium)   123.
long stars   41.
Lo-yang (China)   84, 92-94, 100.
lunar mansions   34-35, 128-129,
    144-145, 164, 195.

heliacal
    rising   88, 109, 115, 146,
    191-192.
    setting   88, 106, 115, 146,
    194-195, 199.

Ma Tuan-lin   41, 140-141, 144.
Maestlin, Michael   174, 186.
Magellanic Clouds   11.
magnitude
   absolute   6, and passim.
   apparent   6, and passim.
Mars   18, 85-87, 104-105, 152, 168.
Maurolyco, Francesco   173.
Mercury   104-105.
meteor   84.
Milky Way   1, 9.
Mira Ceti   50-51, 56, 106.
Moon, brightness of   132-133.
moonlight, interference by 200.
Mugnoz, Jerome   173, 176-177, 179, 185.

Nan-ching (Nanking)   104, 109.
neutral hydrogen
   emission   76.
   absorption   76, and passim.
neutron stars   2, 73.
new stars   1, 14 ff, 40 ff, and passim.
night watches   196.
non-thermal radiation : see synchrotron radiation.
novae,   1, and passim.
   absolute magnitudes of 134, 150.
   brightness of   1.
   cause of   4.
   distribution of   6.
   distribution of nearby 52-53, 150.
   Far Eastern observations of 40 ff.
   frequency of nearby   134, 150.
   light curves of   12, 134, 150.
   range of   151.
   recent   5.
   recurrent   42, 53.
   with deep minima   53, 56.
nuclear burning (fusion)   1.

Observatory, Imperial   26, 29-30.
omens, celestial (see also portents, celestial)   33.
optical
   emission processes   74.
   nebulosities   2, 57, 155, 189, 205.
   searches   180, 202.
oracle bones   19-20.
oriental observations   5, 19,
      25 ff, 40 ff, and passim.
Orion nebula   58.

parsec   9.
Peking,   81, 193.
photometric step   199-200.
Physician, anonymous   196.
planetary conjunctions 168-169.
planetary nebulae   74.
planets, apparent diameters of 133.
Pleiades   145, 165.
Pliny   14.
Polaris   180, 182-184, 194.
polarisation of emission from
   supernova remnants   59, 61, 70-
      77, 158, 170, 188.
portents, celestial (see also
   omens, celestial)   84.
post-novae   150-151.
pre-novae   4.
prognostications, astrological
   25, 28, 33-34, 83, 85, 102-
   104.
proton-proton reaction   1.
Ptolemy, Claudius   120, 172, 179, 181.
Pueblo Indians   154.
pulsars   2, 3, 10, 57, 154, 158,
   160, 213, 216.

qibla (direction of Mecca)   121, 125.

radio
   emission processes   70, 71.
   searches   181.

sources, galactic   2, 57 ff,
    and passim.
rayed stars 40 ff, 148, 175.
recurrent novae 42, 53.
refraction 182.
reign periods 31.
reliability of Chinese obser-
    vations 85-86, 104-106.
remnants of supernovae : see
    supernova remnants.
Renaissance 18, 114, 172.
rock art 153.
runaway stars 2.

St. Gallen (Switzerland)
    122 ff.
Saturn 104-105, 168.
Schüler, Wolfgang 173.
scintillation, ionospheric
    and interplanetary 158.
Scorpio 120.
Sedov solution 207, 208.
sexagesimal cycle 31-32, 114.
sextant 181, 184.
shock waves 10.
sighting tube 169.
Šŏul 191.
sources of new star records
    14 ff.
spectra,
    thermal 57-59, and passim.
    non-thermal 59-61, and
        passim.
spectral index 59, and
        passim.
spiral structure of Galaxy
    217, 218.
star maps, oriental 35-39,
    88-91, 128, 145, 165-167,
    176.
starlight, brightness of 132.
stars,
    apparent diameters of 133.
    evolution of 1-3.
sunspots 91.
supernova remnants 10-11,
    13, 57-82, and passim.
    angular diameter 63 ff.
    catalogue of 64-69.
    distance estimates 75, 76.
    dynamical evolution 63,
        70.

galactic distribution 72.
optical emission 2, 57, 74,
    155, 189, 205.
radio emission 70, 71, 95 ff,
    107, 109, 137-138, 158 ff,
    170-171, 186 ff, 205 ff.
radio evolution 208, 212,
    215, 216.
X-ray emission 72, 73, 158,
    190, 215.
zones 218, 219.
supernovae, 1, and passim.
    cause of 1-3.
    extragalactic 5-6, 9, 11.
    frequency of 7, 208, 218.
    galactic 6-10.
    galactic distribution of 53.
    historical observations of 5,
        11, 13, and passim.
    historical records of 14 ff,
        40 ff, and passim.
    light curves of 9-11, 154.
    searches 9, 191.
    Type I 9-10, 152-154, 163,
        172, 202.
    Type II 9-11, 152-154.
supposed new stars 43-44.
supposed (spurious) supernovae
    44-45, 122.
sweeping stars : see broom stars.
synchrotron radiation 59-61,
    155, 158.

tables, chronological 33, 141.
textual corruption 141, 143.
thermal radiation 57-59.
Tycho Brahe 6, 19, 59, 133, 163,
    172 ff, 191, 196, 199.
Tycho's supernova 6, 59, 172 ff.

ultraviolet radiation 57, 155.

Venus 17, 85-87, 91-92, 104-105,
    150, 168-169, 194-195.
Vietnamese observations 42.
visiting stars : see guest stars.

white dwarfs   2, 4.

X-rays   10.
X-ray remnants   57, 72, 73,
     139, 158, 190.

Zeta Tauri   144 ff.